INDUSTRIAL SAFETY IS GOOD BUSINESS

The DuPont Story

WILLIAM J. MOTTEL
JOSEPH F. LONG
DAVID E. MORRISON

JOHN WILEY & SONS, INC.
New York • Chichester • Weinheim • Brisbane • Singapore • Toronto

Originally published as ISBN 0-442-01842-8

Published simultaneously in Canada.

Library of Congress Cataloging-in-Publication Data

Mottel, William J.
Industrial safety is good business : the DuPont story / William J. Mottel, Joseph F. Long, David E. Morrison.
p. cm.
Includes index.
ISBN 0-471-28628-1
1. Industrial safety --Case Studies. 2. E. I. DuPont de Nemours & Company--Case studies I. Long, Joseph F. II. Morrison, David E. III. Title.
T55.M63 1995
658.4' 08--dc20

95-6533
CIP

Printed in the United States of America.

10 9 8 7 6 5 4 3

Contents

APPENDIXES

Acknowledgements

The authors of this book have drawn primarily on their own experiences, totaling more than 100 years of service in the aggregate, and on their own files for memoranda inherited, sent, and received. But they have also consulted with former colleagues whose specialized knowledge of certain areas of industrial safety have been invaluable in providing the density and immediacy of supportive detail on which necessary generalizations and logical inferences have been based. Chief among them, we wish to thank:

Dr. William P. Bebbington, a former director of the Savannah River Laboratory and author of *History of Du Pont at the Savannah River Plant*.

Arthur Burk, senior safety fellow at the DuPont Company, specializing in process safety management.

James Gentry, a former senior safety consultant at DuPont and, in retirement, an international consultant on safety programs.

Dr. Bruce W. Karrh, DuPont vice-president for Safety, Health and Environmental Affairs.

Donald M. Law, a former editor and public affairs specialist at the Savannah River Plant who provided liaison in gathering materials for the discussions of nuclear safety.

Albert H. Peters, Jr., a former manager of plant facilities and services at the Savannah River Plant.

Dr. Charles F. Reinhardt, director of DuPont's Haskell Laboratory for toxicological studies.

In addition, we would like to express a heartfelt tribute to the several hundred thousand DuPont employees who have worked faithfully and tenaciously over nearly 200 years to build and maintain DuPont's safety record, and to the thousands of safety professionals who have provided vigilant guidance and oversight for this effort. It is their performance that makes possible "the DuPont Story."

Foreword

Among safety professionals, DuPont's outstanding safety record over nearly two centuries has been well known and often envied. DuPont safety specialists have been on the agenda of many safety forums and conferences not only in the United States but in international meetings. They have told various facets of their company's story, and we have listened.

And so I take special pleasure in seeing this story made public in a detailed and comprehensive effort.

Bill Mottel has had a special vantage point in heading up DuPont's safety program after many years spent in industrial management assignments, including a stint as plant manager at the Savannah River Plant, the nation's largest nuclear defense facility. He had also served several years in Europe where safety programs were a central responsibility within the Employee Relations function in

DuPont's European subsidiaries. He also took on the role of an unofficial "safety ambassador," talking safety dynamics to more than 30 audiences here in the United States and in Canada, Switzerland, Germany, England, Hungary and Spain.

Bill is to be congratulated for taking the initiative in his retirement for pulling this story together. He continues to play an active role in the National Safety Council and as a consultant to other industrial firms. His assistants in this venture also bring good experience to bear. Joe Long is the "prototype" plant manager who demonstrated great devotion to safety at several DuPont sites. Dave Morrison wrote frequently on safety subjects in different editorial assignments and advisory capacities.

Their best reward will be a careful reading by safety professionals who are increasingly skilled in adapting proven ideas to their own work environments.

T. C. Gilcrest
President, National Safety Council

Preface

The authors dedicate this study of industrial safety to the many professionals who are concerned with safety on a day-in, day-out basis. These are the people who make the decisions at every level of their nation's industrial organizations, decisions that have immediate consequences on the health and safety of the workers whom they direct or work with shoulder to shoulder. They are also the people who represent the public interest in oversight of the performance of private or governmental institutions.

These professionals include, at the very least:

- *Corporate executives,* who have responsibility for the welfare of their employees just as surely as they have responsibility for the assets, revenues, and earnings of their firms.

- *Corporate line managers,* who carry out management policies and principles as they affect the safety of all employees.
- *Safety engineers, supervisors, and specialists,* who monitor the performance of line operations and serve as resources in ongoing efforts to improve the reliability of equipment and supplies or to make sure that safety factors are not scanted in the necessary efforts to improve cost effectiveness.
- *Proprietors and supervisors* of small commercial establishments, who must at many times wear many hats, only one of which is that of safety oversight.
- *Government regulators,* who must provide the oversight required by laws and regulations while minimizing the inherent adversarial relationships in the public–private interface.
- *Educators and academic research specialists,* who make their own contributions to safety, who train the work forces of tomorrow, and who provide the seminars, conferences, and adult study programs that help managements and rising technicians keep current in a world of rapid change. While this book is not designed as a textbook for classroom, it should prove valuable as a reference. An understanding of and respect for safety is crucial for people preparing to work in the industrial world.
- *Writers, editors and support staffs* in the various communications media, who disseminate and exchange information among all parties in the rapidly changing technologies and must be ever mindful of safety factors.

The material included in this book is not so much intended to break new ground as to bring a broad range of experience to a subject that many practitioners have viewed from a narrow sweep. Nor is it intended as a "how-to" manual, though many helpful hints are offered. It does not supply a long checklist for compliance with OSHA or other regulatory rules. The emphasis is on broad principles pragmatically established. Hopefully the exposition offered here will resonate with the readers' own experiences to give new perspectives in this demanding calling.

The authors' experience largely has been gained in lifetime careers with the DuPont Company, their credentials are stated briefly at the end of the book. They begin their overview of this critical subject with some basic assumptions, which the arguments of this book will demonstrate:

- *Safety is a win–win program.* If it is properly and responsibly applied, everyone wins: members of management, workers, distributors, marketers, civil servants, and, above all, customers and consumers.
- *Safety saves $$$ and maximizes profits.* Every safety program requires additional expenditures on the costs side of the ledger. In the long run, however, these commitments most often avert other, unexpected costs resulting from injuries and damages to people, facilities, and equipment.
- *The cause of safety would be well served by eliminating the term "accident" from the safety vocabulary.* Its use implies that an incident or an injury is simply an "accident," and is not preventable. Therefore, its use has been avoided in the book, except in quoting from sources.
- *Zero incidents should be a universal goal.* To settle for less, however gratifying consistent progress may seem, is to sap the dedicated efforts needed to make safety work.
- *The potential for incidents can be ruthlessly routed out, but risk is always with us.* Consider the risks inherent

> in oceans, floods, tornadoes, and hurricanes, as well as locomotives, trucks, automobiles, saws, hammers, ladders, and chemicals, and even humans themselves. Safety is not the absence of risk; it is freedom from danger and harm. Our goal must be to contain and deal with risks, not to eliminate them entirely.

Finally, some caveats are in order. Although the advice and the review of current DuPont managers have been sought and freely given, this is not a DuPont publication. DuPont management did not provide the initiatives, and the corporation has not subsidized the preparation of this book in any way. The opinions offered in this book are the authors' own.

Summary of Chapters

Chapter 1 provides a brief evaluation of the status of industrial safety today and explains why the authors have elected to focus on the DuPont story as a basis for establishing certain safety principles.

Chapter 2 provides some highlights from the nearly 200-year history of the DuPont Company. Its purpose is to demonstrate the patient but unrelenting determination required across seven generations for the company to achieve a position as a world leader in safety performance. DuPont began as a producer of highly dangerous products necessary to national security and the development of a great nation spanning a vast continent. Explosives were unforgiving of mistakes; so too were the temperatures and the pressures required to process the temperamental materials essential to the chemical manufacturing industry.

Chapter 3 lays down some basic ground rules for establishing a safety program. Emphasis is given to instilling a heightened safety consciousness in employees at every level of the organization—beginning at the very top and permeating every office, production unit, storage center, and shipping and receiving area, as well as every laboratory and the power generation and distribution system.

Constant feedback from every level is required. All employees must be proud of their safety achievements and feel that they have the opportunity to offer constructive inputs to the program. Morale is a necessary building block.

An off-the-job safety program is a vital part of achieving total safety.

Chapter 4 focuses specifically on process safety requirements, demonstrating the complex tissues of responsibility involved and offering a model of analysis and response for every phase of operations. The experience underlying this model is drawn from the chemical processing industry, but the basic model is applicable to industry broadly.

The chapter sets forth a detailed program that complements, integrates, and sometimes leads a total safety program.

Chapter 5 addresses the challenges implicit in offering products that will be safe at every point in the storage, transportation, distribution, merchandising, and ultimate disposal of those products on their way to the user, and in their ultimate disposal. And it details what is required to make sure products are safe in use, whether by other industries or by the ultimate consumers.

Chapter 6 examines the responsibility of an industrial organization to the communities in which it operates its facilities, and recognizes local management's responsibilities for communicating fully and clearly to employee families, local media, community leaders, and educational institutions.

Chapter 7 briefly explores the essential relationships between safety and productivity, both direct and indirect.

Chapter 8 notes how the globalization of industrial production and marketing has added new dimensions to the responsibilities of industrial organizations, and discusses what is required to address those responsibilities across differing cultures and amid rising aspirations.

Chapter 9 described DuPont's extensive experience in the World War II and postwar development of nuclear energy for defense purposes. The company did pioneering work in harnessing awesome physical forces with great potential dangers; yet the safety record achieved in the work was far more outstanding than that of industry in general.

The public has long been content to let the government, scientists, and nuclear pioneers deal with those enormous responsibilities. However, the long-range, constructive use of nuclear energy will require an even greater knowledge of what is involved in making its use thoroughly safe and reliable. Many problems remain to be resolved.

DuPont's positive response to the United States government's urgent request for its assistance, both at Hanford and at Savannah River, testifies to the ability of private industry and government to work together to achieve a common purpose. At its own insistence, DuPont did its part without financial profit.

Chapter 10 observes the interrelationship between traditional safety concerns and the effects upon the environment of modern industrial processes and products. The environmental question is a massive subject in its own right, but industry does recognize and respond to its environmental responsibilities.

Chapter 11 concludes this discussion of the complex issues of industrial safety with some very basic reminders about safety management that can serve as checkpoints in building and maintaining outstanding safety programs.

Four appendixes containing historical and practical information complete the presentation.

1

Safety in a Rapidly Changing Industrial World

Why Another Book

Everyone loves safety.

In the home. On the road. On the waterways. In the air. In the streets. In crowded theaters, nightclubs, and department stores. At the ballparks, on the fairways, in the swimming pools, in the duck blinds, and on the trails where hunters periodically stalk game. Above all, in the workplace.

Yet, unsafe incidents are daily and numerous, and reports of injuries crowd the nightly news reports.

As a social concept, safety ranks right up there with cleanliness and godliness, with motherhood and apple pie. Safety has no foes—only careless and undertrained and undercommitted practitioners.

Figure 1-1. Open construction used in this small production unit is typical of chemical plants. As much as 75 percent of costs in chemical plant construction consists of piping, along with the related valves, pumps, and instrumentation. Complex codes and procedures to assure high quality and safety are detailed in Chapter 4, which deals with process safety mangement.

But, people are *not* equally fond of safety programs.

Shakespeare's most famous character noted 400 years ago that many customs "are more honored in the breach than in the observance." Hamlet was talking specifically about sobriety, but the generalization applies pretty well to safety.

Safety has taken major strides forward during the twentieth century, in all walks of life. This progress has been especially marked in commercial establishments, and particularly in the manufacturing sector where risks tend to be especially great. In the United States during the 1980s, both the annual death rate and the number of deaths on the job dropped 37 percent. The number of people killed at work decreased by 23 percent, to 5714 in 1989 from 7405 in 1980.[1]

But every day sees new episodes recorded on the chalkboard of failure. Despite the improvement, during that decade an average of 17 workers a day were killed on the job. The sense of unrealized goals and objectives is universal.

As a result, practitioners of safety programs faithfully attend major conventions of their craft. They participate in seminars sponsored regularly by academic forums and professional organizations. They monitor; they report; they confer; *they read and they write.*

The authors of this book are part of this fellowship. We believe that what is reported in the following chapters will help it in establishing even stricter goals and winning even greater achievements.

Fortunately there is little in safety information that is proprietary. Safety professionals are generally eager to exchange experiences and information. We all are indebted to each other.

---

[1]Statistics provided by the National Safety Council publication "Accident Facts."

Safety Programs < Safety

Why are safety programs less popular than safety itself? There are several reasons:

Safety programs cost! Although we believe that costs for safety are recovered in improved performance (see Chapter 7), short-term cost reduction programs sometimes result in safety programs getting short shrift.

No responsible executive will knowingly increase risks in order to save a buck here and there. But when managers find it increasingly difficult to meet earnings goals, there is a tendency to defer expenditures thought to be less than critical at the moment. And when there are critical needs elsewhere, high performance people, who might enhance an organization's safety performance, may be assigned to duties outside the safety program.

Careers in safety are often off the main line. Rightly or wrongly, fast track engineers and managers may believe that safety assignments will delay their promotions to higher management levels.

"Tired" safety programs become ho-hum. The virus of MEGO (mine eyes glaze over) is experienced at every level of the work force. MEGO is particularly evident among employees whose duties are routine and hold little promise of increased compensation or advancement.

The antidote for these shortcomings, of course, is imaginative, revitalized, and cost-effective programs driven by exceptional people who care. Said another way: to achieve *zero* injuries we must have all employees from the chairman of the board to the groundskeepers believe in and work toward elimination of all injuries. Central to this is total cooperation and high morale among all company employees.

This message is really the subject of this book.

Safety Program Needs Today

Midway through the 1990s, we face major new challenges. Events of the past 15 years have so drastically altered the conditions under which we conduct business that much of our progress is being undermined. What are some of these factors?

Vast new economic forces have been unleashed by the ending of the Cold War. Rather than bringing universal peace, the political aftermath of the fall of the Communist empire has launched new turbulence. Literally hundreds of ethnic groups seek to establish new nationalistic identities. Vast sums formerly dedicated to defense against large hostile blocs have been made available for constructive developments, but painful realignments of firms and employee groups have led to some market instabilities.

The acceleration of technological and marketing sophistication in the developing countries has created new competitive forces, disrupting established supplier–consumer relationships and placing increasing cost pressures on business firms and their traditional manpower structures. This intensified competition has in turn provoked highly aggressive responses on the part of owners and managements.

Most highly publicized are the waves of downsizing and restructuring. For the time being, profitability is most responsive to the deftness by which many jobs are eliminated, especially in the middle management and support staff structures. There is diminished confidence in the traditional advantages gained by technical innovation in products and processes.

One consequence has been, in some cases, a fracturing of well-established safety structures, and the shifting of safety responsibilities onto people with inadequate experience and training. A corollary consequence has been a deterioration in morale from middle management down to the wage roll employee.

WHY THE EMPHASIS ON DUPONT

We began this task expecting to discuss industrial experience generally, drawing only inferentially on DuPont materials. Early readers of our prospectus, however, urged that we tell "the DuPont story." There appear to be good reasons for doing so:

- DuPont is a very old industrial organization with a continuous history now approaching 200 years in duration.
- Through its long history, DuPont has made and distributed some of the most dangerous materials known to humankind. One by-product has been the accumulation of much safety experience.
- During the recent past, when record keeping has become relatively widespread and uniform, DuPont has held most of the world's industrial safety records. The authors' careers have been part of the effort that made this achievement possible, and we can illuminate the necessary generalizations with considerable hands-on experience.

OUR BASIC CREED

Safety is a powerful leadership tool that enables us to achieve excellence in all we do. High productivity, high quality, high morale, and associated low absenteeism are all products of, and interrelated with, a strong safety program. During the many years of our careers with DuPont, we learned that safety *is* our culture. And we believe today that safety as a practical matter and as a bedrock of corporate ethics must be a primary value, not just a priority in all industrial operations.

These same principles include occupational health and environmental goals. Our goal must be not to hurt anyone knowingly or damage anything in our operations. Safety and ethics are inseparable. We hold that precept dear.

2

The DuPont Safety Story

Beginnings

Many people looking at DuPont's long legal title—which still carries the French name of its founder, Eleuthère Irénée du Pont de Nemours—are not aware that DuPont is the oldest large manufacturing firm in the United States in continuous existence.*

The Jefferson Role

People also are not aware that the original partnership, which was established in 1802, was strongly encouraged

---

*DuPont (with no space) is now used to refer to the company, and will be so used throughout the book in spite of earlier variations (du Pont, Du Pont).

by the deep concern of President Thomas Jefferson about a woeful lack of preparedness in the young republic on the American continent. The country had many patriots in those days, but no Patriot missiles. It even lacked satisfactory ammunition.

The Revolutionary War had demonstrated that there were almost no reliable producers of good quality black powder in the new country. That experience also had shown how extremely hazardous powdermaking was as a commercial enterprise and as an occupation. President Jefferson persuaded the recently emigrated du Pont family to suspend its plans to develop farm lands on the new nation's frontier. He encouraged family members to go into the business of producing high quality blasting powders and military powders (see Fig. 2-1).

American powdermakers had made some acceptable powder during the war for independence although 90 percent of the colonies' needs had been bought from France. By 1800, explosions and British competition had put most domestic mills out of business.

The appeal to the du Pont family was well considered. Eleuthère du Pont had worked under Antoine-Laurent Lavoisier, the great pioneering French chemist. Lavoisier had been entrusted with the small munitions industry under the French revolutionary government. Fortunately, Monsieur du Pont was a quick study in the art of making high quality powder. Soon after du Pont's apprenticeship began, the great Lavoisier was carted off to the guillotine on the grounds that a democratic government had no need of scientists.

Those of us who enjoy all that science and technology have contributed to the American way of life today should be grateful that the democracy of Jefferson rather than the democracy of Robespierre prevailed.

When the DuPont company was founded, the United States was a new society in a new and relatively unex-

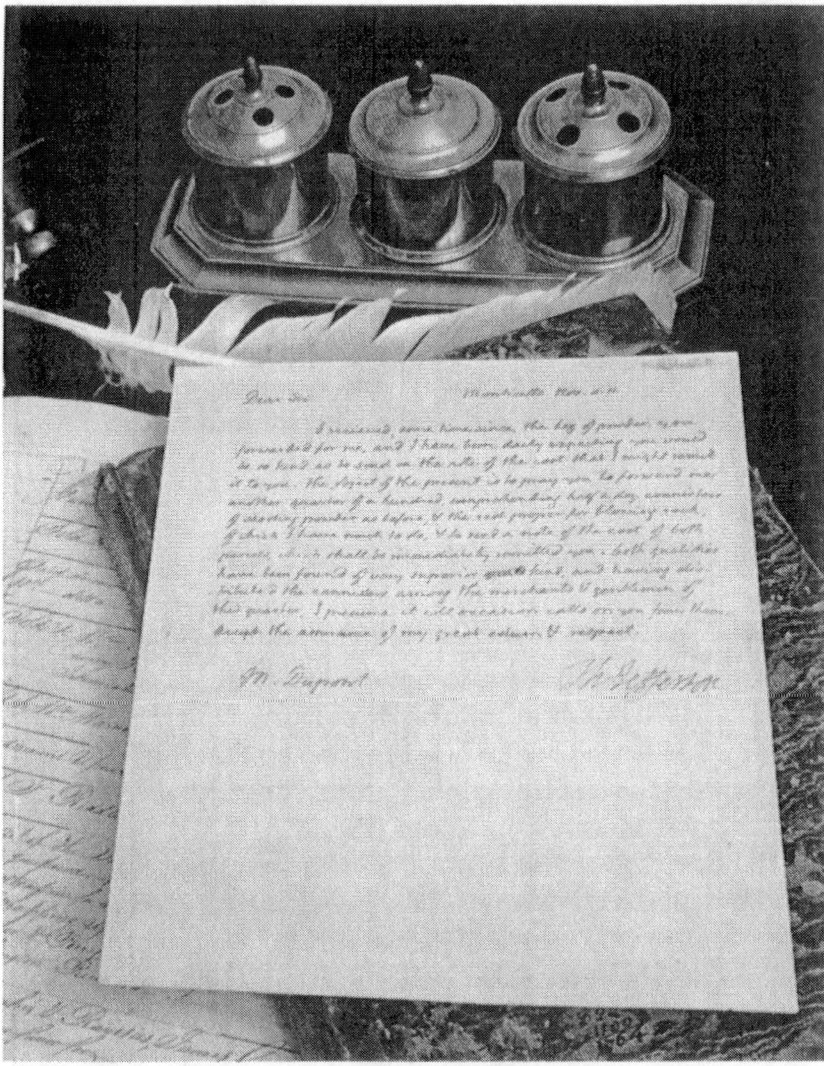

Figure 2-1. Early letter of praise from Thomas Jefferson encouraged du Pont family members to expand powder making operations rather than pursue ambitious land development schemes.

plored world. Black powder was a very dangerous product in an age when manufacturing technology was in its infancy. Within the primitive explosives industry, safety problems, including occasional fatalities, were regarded as inevitable.

The founding family, however, rejected that conventional wisdom. They decreed that people do not have to be hurt. To give vivid demonstration of their commitment, the owners built their homes on hillsides contiguous to the work site, vulnerable to major explosions. The homes of the workers also were nearby, and careless performance could, and sometimes did, jeopardize their families.

Three Walls Funneled Unwanted Blasts

The first three mills were small and separated, to minimize damage if one mill were to have a major incident. Walls were made of heavy granite on three sides, with the fourth wall and a slanted roof made of wood so that the force of any explosion would be vented toward Brandywine Creek and away from other buildings (Fig. 2-2).

The first doctor was hired at $50 for two years' service, evidence of the company's early provision for emergency treatment as well as occupational health.

Company archives also indicate that, in 1806, management gave workers personal instruction on how to roll powder kegs to avoid dropping them. Assuredly, many other safety procedures were discussed in hands-on demonstrations. Copies of written safety rules date from 1811.

Customer response to the high quality powder was gratifying. Production quadrupled within seven years, and a fourth mill was added. The strong growth of the business meant steady expansion, but it also meant constant borrowing of working capital from Philadelphia

Figure 2-2. The Birkenhead Rolling Mills, built between 1822–24 on the pattern established by the very first mills, operated until 1921, the last year in which black powder was made in the DuPont mills on the Brandywine River near Wilmington, Delaware. The picture shows two of the mills as they look today. Safety has been a DuPont watchword since these early mills embodied a design to protect employees. Workers entered buildings only to start or stop operations. Three sides of mills were stone; the fourth side, facing the creek, and the roof were less substantial so that the force of any explosion would be directed away from the powder yard.

banks. The pinch for funds, however, was never allowed to excuse inattention to safety.

Cobblestones Made for Tricky Roads

Conestoga-type wagons delivered powder barrels to schooners waiting at Wilmington docks and fanned out through the countryside (Fig. 2-3). The wagons were not

Figure 2-3. Arrivals and departures of Conestoga wagons carrying DuPont shooting and blasting powders were a community occasion.

permitted to travel in tandem, however, and the prudence of this safety procedure was demonstrated when one wagon headed for the Delaware River went skyward before it reached the docks. Over the years, there were a number of incidents in which the wagon wheels bumped too heavily on the cobblestone streets.

Du Pont Family Took First Risks

Policy dictated that members of the family and supervisors were the ones who started up new machines or initiated new manufacturing processes. A warning message was posted on the doorway of the early powder mills.

> **No employee may enter a new or rebuilt mill until a member of top management has personally operated it.**
> E. I. du Pont

Failures to carry out complex and inflexible safety rules were relatively rare, but the occasional lapse was brutally costly. In June 1815, an explosion killed nine men and caused extensive damage to the mills. These were the first fatalities. Two years later, a worker's carelessness caused the charcoal processing house to burn to the ground and threaten widespread disaster. The du Pont family, including women and children, formed bucket brigades in the fight to control the fire. The only fatality was the aging father of Eleuthère: Pierre du Pont suffered a heart attack brought on by the strenuous effort, and succumbed about three weeks later.

One March day in 1818, an even greater tragedy struck. An explosion wrecked much of the works, killing 40 men. Neither the laws nor the customs of those days committed the company to take notice of the plight of the widows and orphans. But E. I. du Pont pensioned the widows and gave them houses to live in, and undertook the education and medical care of the orphans (Fig. 2-4). There was no full treasury to disburse funds; du Pont paid the cost by renewing his notes at the bank and signing more. Mortgages paid the cost of rebuilding the mills.

We recount these failures to help explain the acute awareness of early company management of the dangers of their chosen business, and to show how early patterns of diligence were constantly intensified. *Attitudes,* so critical to human performance, were forged by hard experience rather than social theory.

WIDOWS AND ORPHANS

July, 1819—Credit to the account $600.00
March, 1820—Debit to the account 600.00
(detail below)

Deposit of $600 sent from Philadelphia for the relief of the sufferers at the explosion of Eleutherian Mills the 19 March 1818 as follows—which sums are passed to their respective credits Petit Ledger.

Widow Bradley	4	children	$105.00
Widow Callagher	2	"	55.00
Widow Bready	2	"	55.00
Widow Reynolds	4	"	105.00
Widow Finigan	2	"	50.00
Clo Bready	3	"	80.00
Tenners	3	"	100.00
P. Quigg			30.00
Hugh Sinch			20.00
			$600.00

Figure 2-4.

A Key Role in the Country's Defense

Military security triggered the powdermaking business in Delaware, with a foresight that proved invaluable in the skirmishes and wars of the nineteenth century. The War of 1812 required a major expansion of productive capacity at the mills. That expansion, however, also set a pattern for the financial stress usually encountered in satisfying defense needs. Revenues increased, but delay in government payments often meant that company management had to take emergency measures to pay company debts. This financial squeeze was felt again during the Mexican War. The Crimean War also increased the demand for gunpowder although the United States was not involved in that conflict.

DuPont was a primary provider of munitions to the Union cause during the Civil War, playing an unpopular role in a border state with a large minority of supporters of the Confederate secession. Stepped-up production resulted in a small number of explosions, with some fatalities. Sabotage was strongly suspected. This experience brought home the need for expanded and intensified security during periods of national crisis.

A Key Role in the Country's Growth

In the decades that followed the founding of the company in 1802, by far the largest outlets for the DuPont blasting powders were the families and individual entrepreneurs who were part of the vast "westering" movements. Early on, the same Conestoga wagons used to transport families to pioneering fields (Fig. 2-5) were used in the distribution of powder. Reliable blasting powders were crucial in building canals and clearing farmland during the expansion to the West.

These early responses to the needs of growth and preparedness also paid off in overseas markets. Today's concern of the American commercial establishment with global enterprise was foreshadowed through much of the nineteenth century.

Transportation methods also changed dramatically through the early decades, placing new constraints on the safe movement of gunpowder to customers:

Figure 2-5. "Westering" movement required many uses of DuPont blasting powders.

When possible, gunpowder was shipped by water. Occasionally railroads were used to transport the product but the hot cinders and sparks emitted by locomotives seriously endangered safety. As a precautionary measure, locomotives were placed behind the train of cars so that cinders and sparks would fly behind the moving train and lessen the unsafe condition. Further, tight roofs and sides were constructed on the cars and damp cloths were placed on the roofs of the railroad cars.[1]

---

[1]Peter B. Peterson, article on "Safety Management at DuPont: A Historical Perspective (1802–1926)," prepared for the *Journal of Safety Research*, October 1986.

Although the early Conestoga wagons continued to be used, mule trains were thought by some to be the most dangerous form of transportation. According to one DuPont historian:

> More than one teamster joined his animals in a sudden trip to Kingdom Come. In May 1854, three du Pont wagons, loaded with four hundred and fifty kegs of powder were rolling through the heart of quiet, orderly Wilmington, jouncing on the cobblestones, when they suddenly blew up, digging a huge hole in the street, knocking down nearby walls and houses and killing two bystanders as well as the three drivers and their eighteen mules.

Actually, such dramatic events were relatively rare considering the total amount of product shipped. But powder blasts had a way of getting people's attention, including that of DuPont owners.

Bigger Bangs without Bigger Bucks? Dynamite Comes on Strong

The westward movement brought considerable affluence and a thorough grounding in competitive experience. DuPont was a major player in the formation of the infamous "gunpowder trust," which established practices later defined as monopolistic. But the hard eye for a buck never persuaded any member of management to skimp on safety.

In fact, an early major falling-out in the family was occasioned by the refusal of the old guard to add dynamite to the product line. Nitroglycerin in any form had been shown to be even more unstable than black powder, and its early use in Europe and Australia had led to many deaths. But younger family members were restive. All had learned the business by serving an apprenticeship in blue collar jobs among the working powdermakers, and they watched nervously as many customers retreated from DuPont as a favored supplier. They saw these customers turn to a succession of U.S. suppliers who began to offer the very powerful dynamite.

The cause of the young "rebels" was led by the most brilliant member of the younger generation. Lammot du Pont had already earned renown by developing a superior brand of powder from Chilean saltpeter (sodium nitrate), as opposed to the Indian saltpeter (potassium nitrate) commonly used. Lammot's impatience grew so great that he decided to leave the family enterprise to start his own business.

In January 1880, Lammot du Pont started construction on the new plant across the Delaware River from Wilmington, at Repaupo, New Jersey. In six months, the Repauno plant[2] developed a production facility that produced a ton of dynamite a day. Aware of the great inherent dangers in this new product, management exhibited even greater concern for safety. Buildings were dispersed even more wisely than at the original company site. High earthen bunkers surrounded many of the units. Severe restrictions were placed on the number of workmen who could congregate in buildings while hazardous processes were going on.

Confident that he could tame the great dangers, Lammot du Pont continued to do research. He was faithful to the family dictum that management must take on extraordinary risks before other employees were subjected to those risks. So he continued to experiment even after his own plant went into production. He paid the ultimate price for this conviction when a runaway reaction in the laboratory went out of control; he was the third family member to die in the dangerous explosives business.

---

[2]A slight change was made in the name of the plant to make it easier to pronounce.

Diversification and the Evolution of Safety Management

The nature of DuPont's business changed radically at the end of the nineteenth century, with diversification into many new product lines. These new lines offered opportunities for innovative technologies, each requiring new safety expertise. Inevitably, safety responsibility became increasingly formalized. A few highlights from this period exemplify the almost "organic" growth that began to take place as early formalized safety and occupational health procedure begat others:

- In 1895–1900, guards were required for moving parts of machinery, and handrails were required for ramps and catwalks. Spiked feet were required on ladder bases. Rubber gloves, aprons, leggings, and special shoes were introduced for those handling chemicals or molten materials. Standard operating procedures were written and circulated.
- In 1907, a first aid handbook was published, describing "temporary surgical measures to be applied in cases of injury pending the arrival of the physician."
- In 1911, a clearinghouse for the study and introduction of all safety devices was established, and operating departments began to set up Prevention of Accident Commissions. A full-time manager, Lewis A. DeBlois, was appointed to head up DuPont's first "safety office." For the next 15 years, his role as a safety manager expanded continuously, and his administrative acumen was largely responsible for developing the safety structure that was to expand as DuPont expanded.

The changes effected by DeBlois extended beyond the Du Pont Corporation. He championed the issue of safety and safety management in many forums. Concurrently with his work at Du Pont during World War I, DeBlois conducted safety inspections of government arsenals and investigated disasters involving munitions for private firms and governmental agencies. After the war, DeBlois was active in safety organizations at both the state and national levels.[3]

- The company began tabulating injuries for analysis. The statistical gathering of performance benchmarks has been invaluable in mapping progress and indicating areas of performance requiring special attention.
- In 1915, periodic physical examinations were initiated for employees. Also injury reports were submitted to corporate headquarters for companywide analysis.
- In 1916, the first separate medical facility was completed at a plant site.

Appendix A provides a further chronology of important safety events, including steps taken during the past 75 years to build a veritable matrix of monitoring and protective procedures.

The Highly Successful Middle Decades

From the end of World War I until the late 1970s, DuPont offered the world many technically sophisticated inventions that took its commercial ventures far outside the

---

[3]Peter B. Peterson, "Lewis A. DeBlois and the Inception of Modern Safety Management at Du Pont (1907–1926)," *Academy of Management, Management History*

original boundaries of chemical manufacturing. Explosives shrank to less than 3% of the company's business.

Synthetic elastomers, man-made fibers, industrial coatings, photosensitive films, polymer resins, and dozens of other extraordinary materials issued from company laboratories. Each research success required similarly bold technological discoveries and engineering developments for manufacture and customer applications. All these innovations, leading eventually to 1700 product lines, strained the boundaries of the company's safety lore.

The caution learned in the early powder mills enabled the company to develop commercial processes for materials most manufacturers thought too risky to undertake. Neoprene synthetic elastomer (the first commercially successful artificial elastomer, perfected by DuPont) was a tiger in a very tight cage. Tetraethyl lead motor fuel additives and chlorofluorocarbon refrigerants (both invented at General Motors) were potentially highly toxic at various stages of their manufacture. The acrylic monomer used to make soft-as-wool Orlon was as explosive as high octane gasoline.

During these decades, however, safety incidents and injuries kept approaching zero as a limit.

Development of the DuPont Safety Record

At the time of this writing, DuPont has been in the safety business for 192 years. Those of us who are now retired continue to be justifiably proud of DuPont's safety record, which is one of the best in industry around the world. Several hundred safety professionals are dedicated to assisting line personnel. We stress the word "assist." Line personnel have always had the ultimate safety responsibility at DuPont.

Figure 2-6 provides some benchmarks that demonstrate what DuPont has been able to achieve:

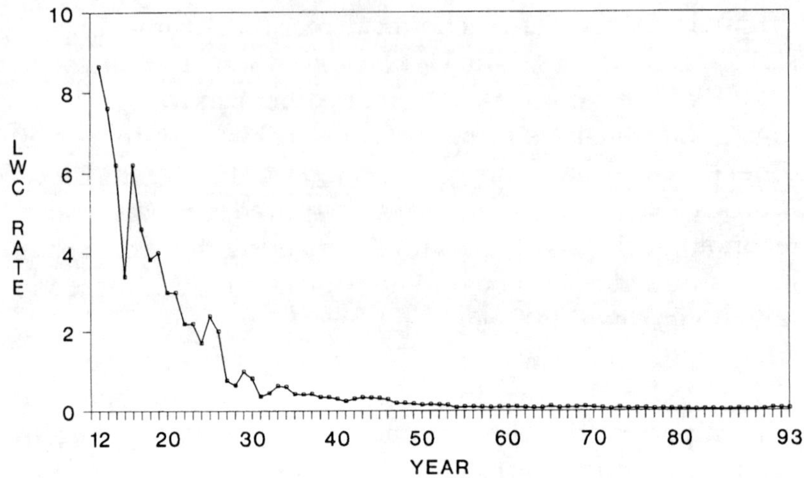

Figure 2-6. DuPont lost workday incidence rate, injuries per 200,000 hours worked, 1912–93.

- The sharp decline in injuries between 1912 and 1931 was due primarily to increasing experience with safety systems and improved technology.
- The sharp peak between 1914 and 1918 reflected hurried wartime production. Explosives were rushed out in huge, improvised plants, with relatively inexperienced workers.
- The minor bump between 1941 and 1945 also represented wartime experience. However, the great volume of experience and accumulated programming kept increases in injuries to a minimum.

The graph does not include records of Conoco, DuPont's energy business, which was acquired early in the 1980s. However, Conoco continues to be the leader in this field.

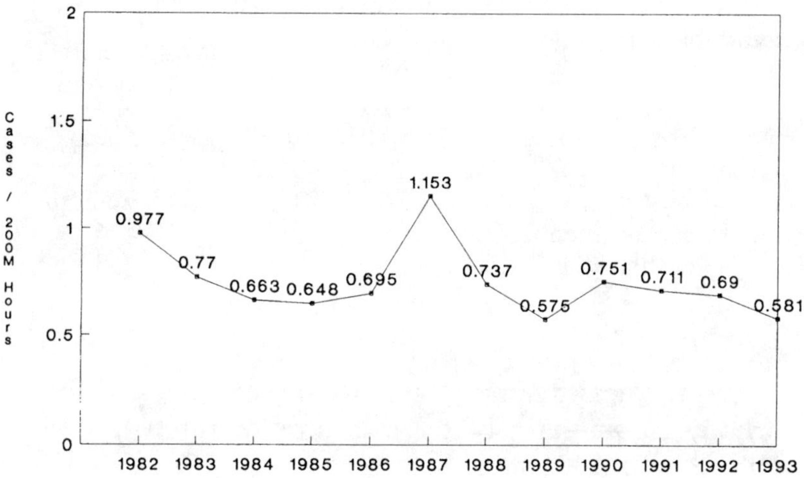

Figure 2-7. DuPont Chemicals and Specialties sectors, worldwide experience.

Overall Lesson

This good record came about through the intensity of management commitment, an ongoing training program that reaches every individual throughout DuPont's worldwide organization, and the ability to upgrade programs constantly in the light of experience.

Figure 2-7 gives us a close look at what approximates "near zero." Notice that all-time lows were reached in 1985 and 1989. The rise in 1986–87, however, was deeply disturbing.

We need to examine the latter phenomenon. Although the increase at first only looked like a blip on the chart, it clearly indicated that more people were getting hurt, and somewhere there was a letdown in the company's program. The company was very concerned—but more on this later in the chapter. First, let us continue to look at the good side (see Fig. 2-8).

To meet the company goal of zero injuries, DuPont now focuses more on total recordable cases in OSHA termi-

Figure 2-8. 1992 National Safety Council Industry and Chemical Rates

	LWC	Total Rec.
Chemical	0.50	3.50
All Industry	2.10	9.10
DuPont C&S*	0.047	0.609
Times safer than chemical industry	11	6
Times safer than all industry	45	15

*1993 Data

nology—medical treatment, restricted-workday, and lost-workday cases—as a better indicator of safety performance than traditional measures.

These figures show that DuPont performance in lost-work-time cases, which represent the more serious injuries, is *11 times safer than the rate for the chemical industry taken altogether, and 45 times safer than the rate for all industry.*

Safety as Vision and Mission: The Contemporary Setting

We have been asked many times what the DuPont secret is. It is not magic or voodoo. It is a *state of mind,* a point of view *shared by all employees.* The following slogans are part of everyone's training:

- Safety is our heritage.
- Safety is good business.
- Safety makes us totally credible.

DuPont history tells us that taking the pains to be safe makes a company more competitive in the long run. As noted in the preface, safe operations are efficient operations; safety, productivity, and high quality go hand in hand. Safety is a powerful leadership tool that enables us to achieve excellence in all we do. A sincere concern for people is nothing to be ashamed of or to be put in a cost–benefit equation. DuPont is convinced that attaining mutual confidence, respect, and trust from the top down is fundamental to a superior operation. For those readers who have had experience throughout the chemical industry, it would require nearly $2 billion a year in additional sales to pay the cost. We further explore the relationships of safety and productivity in Chapter 7.

Acknowledging that safety is good business, DuPont proceeds with the conviction that safety is rooted in a sincere concern for the individual employee and his or her family, a strong commitment from the top, and hard work at all levels.

In this definition, commitment means more than good intentions, or the setting up of standard procedures. *It means the constant, daily, dedicated—let us say passionate—involvement of every level of management.* It also means the understanding of every employee from the day she or he is hired that safe performance is more than a goal; *safety is a condition of continued employment.*

In the 1990s, both DuPont ownership and facilities are widely dispersed. But the company's chief *executive* officer is still the company's chief *safety* officer. Concern must start at the very top and work its way down with equal force through every operational level. While we worked, we had a slogan on our desks that every visitor looked straight at (see box). And we were gratified on our visits to more than 100 company sites to find that same motto on many office and locker room walls throughout the organization.

YOU WILL ACHIEVE
THE LEVEL OF SAFETY
THAT YOU DEMONSTRATE
YOU WANT TO ACHIEVE.

Figure 2-9 shows how safety is structured within DuPont.

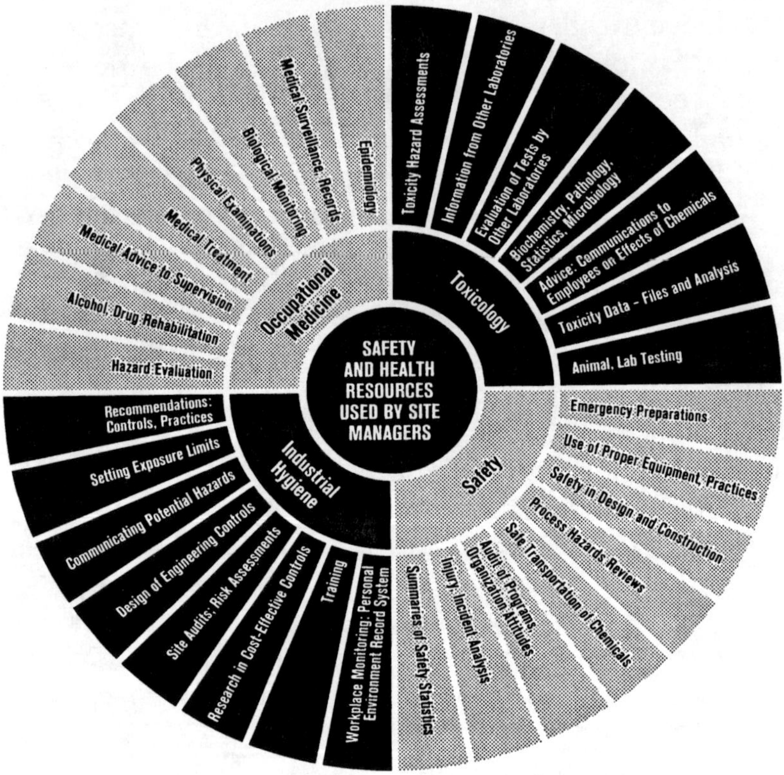

Figure 2-9.

In the company's operational meetings, from the Office of the Chairman down through daily "rackups" on manufacturing plants, the first item of business is always safety performance. Any lost-time, on-the-job injury or illness involving a DuPont employee anywhere in the world must be reported promptly, usually within 24 hours.

Safety and Health Principles

Behind this commitment is a series of principles guiding management in its stewardship of the safety function:

- All injuries and occupational illnesses are preventable.
- Management responsibility for preventing injuries and illnesses extends through first line supervision.
- Safety is a condition of employment.
- Safety must be a part of each employee's training.
- All deficiencies revealed by audits must be corrected promptly.
- Prevention must be the reaction to *incidents,* without waiting for *injuries.*
- It's *good business* to prevent injuries and illnesses.
- Off-job safety is as important as on-job safety.
- People are the most important element of a safety and occupational health program.
- Each and every individual has a responsibility to work safely with proper regard for the safety of others.

Managers and executives are regularly reminded that the safety records of their operations are considered *first* in the evaluation of their performance. This is true both for promotions and for yearly incentive compensation awards. Although contributions to profitability are extremely important, managers understand that business strategies will never excuse their taking their eyes off safety and environmental control.

Within operational units, *every employee is required to report every safety incident,* no matter how trivial it may seem at the time. Such reporting may help prevent a similar pattern of failure or laxity when the consequences might be serious indeed. Safety concerns do not stop at the property line.

DuPont Safety Policy

DuPont's worldwide corporate safety policy statement makes its position clear:

- We will comply with all laws and regulations.
- We will routinely review our procedures to upgrade *beyond legal requirements.* We will be proactive.
- We will ensure that each product can be made, used, handled, distributed, and disposed of safely—otherwise we won't make it.
- We will inform employees and the public about our products and workplace chemicals.
- We will provide leadership to communities in response to emergencies.

Occupational Health and Product Safety

In the period between the great world wars, DuPont became increasingly aware that skills used in the anticipation and the avoidance of injuries needed to be extended to the realm of occupational health.

The establishment of the Haskell Laboratory for Toxicology and Industrial Medicine in 1935 was an outgrowth of DuPont's pledge that the company will not offer for sale any product that cannot be produced, or handled, or used safely (see Chapter 5).

In a sense, the need for the laboratory grew out of the company's broadened program of diversification following World War I. The dangers of many new chemicals were often less obvious than those of explosive and flammable materials. When it was founded, the laboratory was a pioneering venture in industry. Its early work focused almost entirely on new developmental products, whereas in subsequent decades the laboratory investigated health hazards that could conceivably crop up in the working conditions of its employees.

The Zero Concept

At DuPont, all supervision agrees that the fundamental safety goal is zero injuries. This goal was stated in management directives as early as 1915, and it continues to be basic today.

The company has critics who sometimes say that because zero is rarely achieved, it is an unrealistic goal. They suggest that more is to be gained by emphasizing small improvements over the best previous record. But DuPont believes that the company can get the maximum benefit by pursuing both points of view. Since it began keeping records in 1912 (again see Fig. 2-6), the company has decreased its lost-time injuries by a factor of over 250. There are sites with hundreds of employees that have gone over 25 years without a lost-time injury.

A zero goal encourages all employees to be more vigilant, more demanding of themselves, their subordinates, and even their superiors. A zero goal forbids complacency. It signals management's refusal to relax just because improvements have been conscientiously made.

DuPont's Use of Safety Data

There is great merit in the use of statistical benchmarks. Management uses quantitative data to monitor injuries, property loss, transportation incidents and crashes, releases and spills, and off-the-job injuries. These data serve several purposes:

- They indicate progress.
- They spotlight geographical sites with special problems.
- They focus on weak points in the organizational structure.

- They identify processes that need tightening.
- They promote pride in performance and teamwork.

Blips on the Chart

Now let us return, as promised, to the problem years of 1986 and 1987. Then, the statistical benchmarks helped to alert DuPont to potential problems of some significance.

The year 1985 represented an early peak year in DuPont's program to streamline administration and support services. It had become clear that the company had to get its numbers down to remain competitive in the global marketplace. To accomplish this as comfortably as possible, and to maintain a youthful and energetic employee population, DuPont offered very attractive early retirement options to those people who might wish to retire early. This pattern has prevailed during even greater downsizing in the 1990s.

The response was gratifying. During the mid-1980s nearly 40,000 employees left DuPont, but the downsizing also left the company with some very real gaps in capability. When it cut fat, it also cut some muscle. Some units said goodbye to as many as 25 percent of their people. At the same time, the traditional eight levels of supervision in the operating departments dropped to four levels or even three. Many of the company's supervisory employees took on additional responsibilities and tasks. Managers diffused their monitoring efforts as they were assigned spans of control, adding anywhere from two to twenty people to those they supervised.

In the process, DuPont was forced to give up its traditional management belt-and-suspenders approach, that is, backing up every system. No longer could there always be an ingrained system of check, recheck, and crosscheck.

One result of this streamlining was that the old system

of driving, and enforcing, safety performance mainly from the top did not work quite so well as it did in the past.

The problem was compounded by fundamental changes in the company's corporate culture. Great emphasis was placed on partnering with customers, backed by closer attention to product quality through improved work performance at every level.

These changes of emphasis were laudable and greatly enhanced DuPont's marketing efforts across the board.

In carrying out these modified objectives, there was never any intent to downgrade safety. But management was too confident of the inherent role of safety in the corporate culture. Managers relied on the long safety tradition and believed that momentum would carry it through while they focused their main attention on executing the new programs.

Management was wrong—very wrong—in this.

For example, it had been a policy that even production line workers, in certain instances, could shut down an operation on their own if they perceived a safety problem. And, another example, managers traditionally felt empowered to either repair or replace machinery immediately if safety was considered to be an issue.

But with increasing emphasis on cost-effectiveness, some managers gave priority to uninterrupted production runs. They hesitated to correct minor safety problems if doing so would require costly down time. Instead of insisting on safety at all costs, some work groups began to consider "reasonable" safety performance an acceptable goal. They began to take shortcuts.

Although such a variation in emphasis may seem small, it amounted to a fundamental change in a company where safety management had always aimed for *zero* injuries.

The downslide in performance that began to show up in the company's safety statistics in 1986 continued into

1987. There was a temptation in one department showing a small upsurge in incidents to attribute the problems to new employees. Closer examination, however, showed that veteran employees were making the mistakes, in part because experienced supervision had been reassigned to train new employees.

Although DuPont's overall safety performance was still one of the best in manufacturing anywhere in the world, senior management considered the situation alarming. Immediate action was taken to help turn things around.

First, *senior management made it clear that safety remained a basic company value and that the new culture would have to adapt to basic safety needs.* Management made especially clear that there could be no compromise with safety standards or relaxation in the goal of zero injuries.

Second, *the company made it clear that it now relied on individual employees to be self-managed and self-motivated where safety was concerned.* That had always been the policy, but very tight safety supervision had masked the fact that many employees relied on their supervision to ride herd. Now all employees were reminded that they would not be judged simply on their ability to follow the lead of others in carrying out safety programs.

Safety Roles for Employees

The following employee roles have been defined:

- Identifying and communicating hazards.
- Developing safe operating procedures.
- Developing training procedures and programs.
- Investigating incidents and injuries.
- Participating in workplace health and safety program activities, including health and safety committees.

The results of these efforts are showing up in the 1990s in better on-the-job safety performance, even though downsizing continues to reduce employee numbers throughout the organization.

A remaining question in the minds of many readers may be: What programs are required on a continuing basis to sustain this commitment and to accelerate improvements toward defined goals?

Let us mention a few of DuPont's supportive endeavors.

In-house Support

In-house, the company provides the following:

- Constant reminders to all levels of management that safety gets first priority as a basic value in DuPont.
- Ongoing training of safety professionals to provide certain expertise needed by line management.
- Continuous training of *all* personnel in safety procedures and practices.
- Regular but not predictable safety audits.
- Well-structured two-way communications.
- In production operations, daily time spent on the floor by the top manager or his or her designate.

Operational Basics

The following objectives are kept before all management people:

- Define accountability.
- Establish clear performance standards.

- Measure performance.
- Take appropriate action.

In setting clear-cut performance standards, rules are essential. But they provide a special problem because people tend to get bored with them. And human beings get careless. Industrial people (outside DuPont as well as inside) think they are all on board; yet, every week seems to bring new headlines about major incidents, usually caused by human error. The best technology will not save us if people sleep below decks while on the job, or fail to concentrate hard on routine inspections, such as looking for cracks around the rivets that hold the skins of airliners together.

Off-the-Job Safety Programs

Off-the-job safety is a worthy goal in its own right. Encouraging employees to be safe away from work helps to preserve skills that could be temporarily or even permanently lost. It has the additional value of keeping absenteeism low.

Being careful *everywhere* is a good mindset. One's concentration on the job is likely to be enhanced by one's concentration off the job. And the plant or the office is a healthier, safer environment because of it. This subject is explored in detail in Chapter 3.

Networking with Other Organizations

Through most of its history, DuPont has pushed the frontiers of safety knowledge and performance, but increasingly it has carried its message to industrial safety organizations and learned much from them in turn. This is especially true of the National Safety Council at large and of its special committees and subgroups.

DuPont also has worked intensively with the Chemical Manufacturers Association and with the railroad industry to find ways to ensure the safe transportation of hazardous materials. In 1971 DuPont helped set up an industrywide transportation incident and response system that was operated by the Manufacturing Chemists Association (MCA), a forerunner of the present CMA.

In 1978, this effort was expanded into a formal joint effort, the RHYTHM (Remember How You Treat Hazardous Materials) program, as a response to an alarming increase in transportation incidents. Many of these incidents involved the transportation of hazardous materials over railway lines where maintenance of railbeds had faltered as a result of deteriorating economic conditions in the 1960s and 1970s. RHYTHM has had noticeable success in reducing the threats to public safety from this source. (See Chapter 5.)

Relations with governmental oversight bureaus and administrations have not been so sanguine. Unfortunately, it is inherent in a governmental organization such as OSHA (the Occupational Safety and Health Administration) to pursue its mission from a regulatory or rulemaking stance, rather than by performance. Often the standards OSHA is charged with enforcing are less demanding than DuPont's own self-imposed performance objectives and design standards. Consequently DuPont tends to think of OSHA requirements as checkpoints, to be carefully observed in good faith, rather than as blueprints for achieving even better statistics. In some cases DuPont practices have helped to shape OSHA guidelines.

Government—Regulations vs. Guidelines. DuPont believes, and the authors believe, that guidelines with performance-based recognitions and penalties are a better alternative to government involvement in industrial safety.

On the Debit Side

The rare failure still can have very tragic consequences. A major fire at Du Pont's Louisville, Kentucky neoprene plant in 1965 resulted in loss of life as well as major damage to the plant. An explosion at an explosives plant in Carney's Point, New Jersey in 1976 resulted in fatalities. An explosion at an explosives plant at Goiabal in Brazil in 1983 also caused several fatalities.

In 1989 DuPont decided to withdraw altogether from its historic explosives production in the United States.

Long-Term Hazards

Like other producers of chemicals, DuPont has been haunted by the gradual discovery during the past 25 years that exposure to certain chemicals, kept within guidelines that showed no immediate consequences, can result in physical impairment one, two, or three decades after the exposure.

An example of this discovery—which was highlighted in the 1980s but was becoming evident in preceding decades—can be seen in DuPont's dyes business. During World War I, DuPont invested $27 million—the equivalent of $600 million in 1994 dollars—to develop a stable of organic dyes. These substances were needed to fill the gap left by the Germans when their normal business operations were disrupted. Over a period of several decades, an epidemiologically significant number of employees, including managers as well as operational employees, developed cancer of the bladder.

Epidemiological studies today are a major part of the effort at the Haskell Laboratory (see Chapter 5), but many enigmas remain unsolved. At one time the Ames test, named for its developer Bruce Ames, was used to detect cancer potential in the belief that mutagenicity was a reliable predictor of carcinogenicity. Today even Bruce Ames is no longer confident of the total reliability of these correlations.

To sum up, safety continues to be an ongoing process of discovery and disciplined application of findings. But DuPont, and we who administered its programs for several decades, are confident that a broad-based policy is possible, useful, and efficacious.

3

General Application of Safety Principles

Establishing a Safety Program

Old English formula for creating a beautiful lawn: "Plant good seed and mow for 500 years."

That formula has been successfully followed but it is not the only way to go. If it were, the American colonies, less than 500 hundred years old, would only now be closing in on beautiful lawns.

The Vanderbilts and other creators of great country estates in the United States, enamored of the greensward surrounding great English country homes, tried to shorten the formula by importing shiploads of sod scalped from rural English lawns. Today, in an age of hybrid seeds, crop protection chemicals, and rototillers, that kind of effort would not be considered "cost-effective."

The preceding chapter describes very briefly the Du-

Figure 3-1. Mechanics in special protective suits change a filter in a chemical intermediate supply line.

Pont safety tradition, which has been building over two centuries. Like today's landscapers, however, who can create lovely lawns in a matter of weeks, industrial safety managers do not need two centuries of heritage. They can draw up and establish safety programs very promptly if they follow certain well-established principles.

Implementation from the Top—But Carried All the Way Down

DuPont found that the necessary commitment to safety on the part of operating people is unlikely unless that commitment begins at the top level of management. And it must be carried out on a daily basis at every other level of management and down through every nonsupervisory employee. (See Fig. 3-2.)

Managers, operators, supervisors, operators, and me-

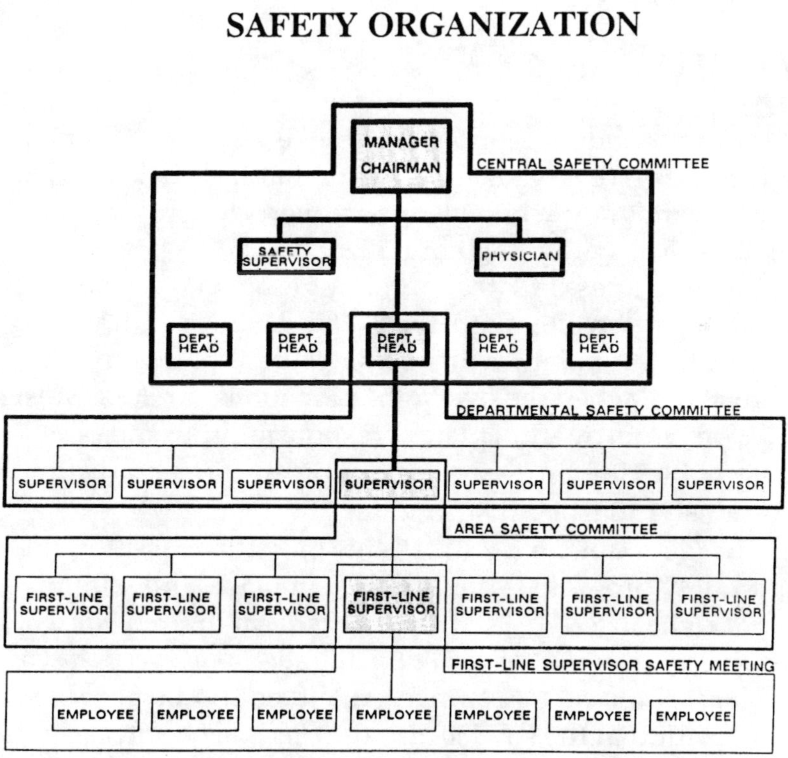

Figure 3-2. This simplified safety organization chart is a good starter for setting up a safety structure. It can be expanded for larger organizations, or duties can be combined for small units.

chanics must continually make decisions regarding conflicting priorities in areas such as production, quality, cost, personnel, and security, as well as safety, occupational health, and the environment. In effect, they must constantly make risk/cost–benefit decisions on the spot, often without maximum deliberation or guidance.

Occasionally other priorities will get placed ahead of safety. This is a highly undesirable situation that will increase the frequency of process incidents.

In short, management must reaffirm the value of safety as a top priority, not only by directive but by example on a daily basis.

The CEO Takes Charge

Almost without exception, the chief executive officers of businesses of every size are aware that the buck stops at the top, and they acknowledge an overriding responsibility for safety within their organizations.

Yet the success of safety programs is uneven at best, and the fault often lies with the CEO. There is a fine line between delegation, which is absolutely necessary, and dereliction, when the oversight function is careless. Most CEOs have an overfull plate of responsibilities, and potential troubles lurk in every phase of production, marketing, administration, and distribution.

The CEO can cope with all this only if a well-thought-out safety program is devised, put in place, and administered rigorously. *A passionate commitment to safety must be demonstrated at the top and be equally evident at every level of the line organization.* Company manuals should record in bold print that the CEO is also the *chief safety officer*.

Periodic staff meetings, usually held at the beginning of every week, should begin with a review of safety program developments and maintenance in every division, including the reporting of all incidents deserving attention. Managers of administrative units must be equally attentive and forceful: although major incidents are less frequent in offices, even there secretaries have been wounded by unstabilized file cabinets and improperly grounded appliances and business machines.

The CEO's personal impact cannot be simply signaled from his or her office. The CEO must talk and write about safety and when possible be seen where the work is done—on the plant, in the laboratory, or in the offices. A CEO should welcome the opportunity to talk to operators, mechanics, millwrights, and filing clerks about work problems, work improvement, prevention of incidents, and the proper use of safety equipment. In large, geographically distributed organizations, the site manager must act as the CEO's surrogate.

There is another important payback in making a walking tour of operations. One can get much better feedback in a regular, person-to-person tour than one will ever get sitting at a desk, no matter how important one's office duties are.

W. Edwards Deming—whose hard-won fame as a quality control expert, first in Japan and tardily back home, testified to the importance of close, direct observation—emphasized that improving quality sets off a chain reaction. That reaction, he was able to demonstrate, increases productivity, reduces prices, increases market share, and ultimately leads to more jobs.

Deming's dedication to "measurement" in quality control is analogous to the safety professional's passionate pursuit of a "zero incidents" goal. But Deming also found a close working relationship with mill employees to be as fruitful as the study of statistics in his office.

Many plant managers have found it effective to "walk and talk"; that is, to take a stroll the first hour of each regular workday, and to converse with selected individual employees when this does not require interruption of their working duties.

The Corporate Safety Division

In large organizations, a corporate safety division generally is staffed with safety professionals who keep abreast of government regulations, develop informational resources, counsel safety units at local sites, and provide oversight for corporate safety performance.

These safety professionals, whether congregated on headquarters staffs or distributed at sites, are an invaluable resource in all aspects of safety programs. However, too many industrial organizations are content to assign the safety responsibility to solid performers who, for one reason or another, are off the fast track for promotions.

Some of these people can be depended upon to give very strong performances; but the cadre of safety specialists also should include *highly promotable people,* who can benefit greatly by working full-time for a defined period where they have a daily view of the significance of safety operations in the broad scheme of things. Important as such assignments are, however, there is an all-too-frequent risk involved. High potential people must not be overlooked or have their career paths shortcircuited by being out of the eye of line management.

The principal responsibility for safety, however, remains with the line organization.

The "Plant Manager"

As noted, in organizations with multisite operations, the division manager, department head, site manager, plant manager, or laboratory director serves as the viceroy of the CEO. In the safety program, this individual also oversees and supervises activity at that site with the same passionate conviction that the CEO demonstrates in the corporate structure. His or her responsibility usually is executed through lower tiers of management, including first line supervision.

The Central Safety Committee

The local manager also carries out the safety function as chairman of the safety committee at its periodic meetings, usually monthly.

Depending on the size of the organization, the central safety committee normally includes division or section managers, superintendents, and the Safety Engineer. The committee also should include one or more supervisors at the next lower level—for feedback purposes, for implementation, and as part of their own managerial development programs. The inclusion of wage roll employees, particularly on action committees, has proved to be very effective.

It is the Safety Committee's responsibility to make sure that all safety aids are in place, that the safety messages are communicated, that the company's safety policies and rules are published and evident, and that employee training programs incorporate safety instruction and indoctrination. The committee also initiates and oversees the special programs that help to keep safety practices up to date in areas of new knowledge, new technology, and overall safety program enforcement. A series of special programs can make safety consciousness fresh and interesting.

Figures 3-3 and 3-4 show safety signs and protective clothing, both of which are concerns of the safety committee. (See also Figure 3-1.)

Figure 3-3. Large, visible signs are constant reminders to all personnel to be vigilant.

SAFETY
SHOWER

CHESTNUT RUN
PLAZA
EMERGENCY
RESPONSE
TEAM

Figure 3-3. (cont.)

The Safety Engineer

One or more safety engineers carry out the following functions:

- Daily site monitoring.
- Working with supervision at the levels closest to operations, and reporting potential problems to higher levels of supervision.
- Keeping track of applicable OSHA regulations to ensure compliance.

Figure 3-4. For protection against the remote possibility of a leak, the chemical operator who disconnects this tank car must wear a full acid suit made of Neoprene synthetic rubber, with a full-face breathing air mask and an independent air supply.

- Suggesting and directing special programs.
- Keeping all safety data, including data on incidents, injuries, and progress in meeting award requirements, such as days when qualifications can be reached.

- Helping to publicize award programs.
- Participating in safety audits, in cooperation with designated audit groups.
- Assisting the training of individuals and small groups on safety fundamentals, and monitoring their actions for approval or correction.
- Making sure that safety protection equipment is the very best available and is used without fail.

Monitoring Environmental Performance

Automatic instruments perform chemical analyses, and recorders indicate temperatures, pressures, off-gas purges, flows, and so on. Environmental engineers or safety engineers performing a dual function are responsible for checking data and making corrections. Official record keeping may be handled by the safety engineer.

Safety Audits

Audits are done by many groups. These audits may include:

- Governmental audits.
- Audits by higher management.
- Central safety committee audits.
- Daily inspection by the site manager.
- Audits by area supervision, sometimes in connection with designated wage roll members.

Audits may be comprehensive or, at times, may focus on such specific details as "locking out and tagging" during repair procedures (Fig. 3-5). As a practical matter, audits usually should combine both safety and house-

Figure 3-5. Personal protection and equipment safety go hand in hand whenever maintenance work is performed. This employee, wearing protective gear specified for the job, works on valves that have been "locked and tagged out." The tags, signed and in place, signify that the system has been made inoperative and poses no danger to the employee.

keeping, as these two considerations generally go hand in hand.

Incident Reports

Lost-time injuries are almost universally used as a measure of safety performance. And the history of events leading to such incidents helps to sharpen prevention efforts to make sure there is no recurrence of those events.

But there is much to be gained in putting equal supervisory effort and emphasis on *incidents that do not result in injury*. These incidents should be thoroughly publicized on the site. An increase in the number of incidents should be given priority as a topic at safety meetings, to stimulate closer oversight and, where feasible, to initiate special procedures to build safety awareness. Otherwise, the increased frequency of incidents will inevitably lead to the injuries everyone is trying to prevent.

Investigation Procedures

When incidents do result in injury, the site manager generally will select or approve the selection of a group to investigate the circumstances and make both written and oral reports to the central safety committee. The investigative group should include supervision from the work area where the incident occurred, a safety engineer, and one or more technical specialists. Where possible, the injured employee should participate in the investigation. The final report should not only establish the causes of failure, but also recommend actions to prevent any recurrence. News of lost-time injuries should be immediately published throughout the organization. Such reports are regularly reviewed by safety committees at other sites; they can be discussed to advantage as means of avoiding injuries at these other locations.

Records

Records of injuries are kept centrally for worldwide operations as well as at individual sites. The records facilitate measuring current performance against past performance and against the performance of other work units. Most people are competitive and like to excel in activities they share with their peers. Safety is no exception. Informational bulletins should make effective use of comparative statistics, graphs, and photos to stimulate employee interest and the desire to excel.

Many individual sites benefit from placing billboards at entrances to keep a running score of hours worked safely. Sometimes the statistics are augmented by a green light as safety hours accumulate, or conversely by a red light to signal an injury—which, of course, records poor performance. Over a period of time, there is evidence that the stimulation of a good safety performance correlates with improved productivity and quality.

Training

Properly trained employees are an absolute requirement for keeping complex process equipment and machinery on the prescribed path to safe operation. All other key elements of process hazards management can be in place, but without properly trained and thoroughly committed personnel, the chances of safe process operation are greatly diminished. (Figure 3-6 shows a plant firefighting squad in a training exercise.)

The great catastrophes of recent years—Flixborough, Seveso, Bhopal, Chernobyl, Sandoz—all owe their occurrence to failures of people to carry out prescribed proce-

Figure 3-6. Training of plant personnel in proper techniques for fighting fires is a major defense against potential catastrophes.

dures that could have prevented or greatly mitigated the consequences of process irregularities. (These incidents are discussed further in Chapter 4.) Usually people fail because they are inadequately trained, bored, tired, distracted by personal problems, or just plain forgetful.

In training and orientation, every employee must understand that rigid adherence to all safety rules is the first condition of continued employment. Although safety is a first priority for all management personnel, each wage and salary employee must recognize her or his responsibility and be encouraged to take initiatives—and must be rewarded for exceptional contributions.

Awards and Recognition

For management people, safety performance must be a major factor in performance reviews, in ratings of promotion potential, and in determination of annual incentive awards.

For the larger employee population, awards to individuals within the group based on group performance, preferably in the form of useful articles, may be used as a symbol of recognition. The major function of individual awards is to generate peer pressure so that each worker will expect to be "his brother's keeper," recognizing his duty to help coworkers to be safe. Such awards also allow workers to show an object of tangible value to their families that indicates that their groups have excelled in a major aspect of performance.

Although such tangible rewards, given at major safety milestones, are the most publicized, other types of rewards can be used. A letter from a site manager to an individual who has worked many years in a safe manner is treasured. Small cash awards may be treasured for the recognition much more than for their monetary value.

In large work organizations, special ceremonies with displays of awards can be arranged on the basis of millions of work-hours without a lost-time injury. But rewards must be attainable in a reasonable time span; thus a favored recognition criterion may be to award each member at a site for each 365-day injury-free period.

Special Programs

These programs largely arise from the initiative and ingenuity of the local safety committee, and reflect the local culture. Rather than attempt to define them, let us offer just a couple of examples:

Silver Dollar Campaign. In this effort, the safety engineer or other appointed individual may "hide" a silver dollar in a spot that needs attention. The finder keeps the dollar and is recognized for finding it.

Safety Fair. The basic principle is that people like having their families see and recognize special work they have done. Frequently a safety fair may be part of an employee open house. Families, usually with no children under 12, are invited to walk through the work area at a set time. Guides are provided. As an example of individual accomplishment, a mechanic in the shop may demonstrate his or her particular improvement. At one plant, such "homespun technology" included:

- A special buggy designed both to transport equipment needing repair and to provide fixtures needed in the repair.
- A cabinet with graphics showing varied sizes and types of gaskets so that mechanics could have a clear idea of the gaskets that were discarded and replaced.

Feedback from the Bottom Up—and Encouraged All the Way Up

Management has found that the best approach is to "get the wage roll on our side." All personnel must accept without question that a safe, clean workplace benefits all people on the site.

Window dressing will not do. No clean up, dress up, or straighten up should be necessary for visitors, auditors, or inspection. The need for a daily personal look by the plant manager or site manager, with comments and discussions on the spot, cannot be overemphasized. People like to participate and to talk to the person in charge. Unfortunately they are not always so eager to take on responsibility. The top manager at the site should ask for everyone's help to maintain clean, safe, trouble-free operations so that they all can go on working there and enjoy doing so.

Of course, management, including the top manager, must show genuine interest in the work force—they do not want to be treated as numbers. When people know they are valued and respected, they will be forthcoming with their own observations. Including wage roll representation may have the good effect of having workers at every level participate in writing and rewriting policies and procedures, to make them more meaningful to the whole work force.

Morale, an Essential Ingredient

A good safety program can be a real cornerstone in building morale because the program itself is concrete evidence that management is willing to spend time and money to prevent employees from getting hurt. Conversely, good morale is an absolute necessity in building and maintaining the discipline required for process safety management.

Morale may be defined as:

- The mental and emotional sense of personal well-being and job satisfaction that underlies the commitment of all individuals to the tasks expected of them by their group and their management; or,
- A sense of common purpose shared with each member of a group that engenders confidence in the face of difficulties and setbacks.

Experience shows that a group with good morale generally performs its tasks—including safety, production, and innovation—in the most effective manner. However, in the present period of corporate consolidations and downsizing, and of computer-based techniques designed to reduce the need for intermediate levels of management, maintaining good morale in the total enterprise is difficult.

The following general principles are relevant to any discussion of morale in industrial organizations:

- A consistent management policy encouraging individual development helps to improve or sustain morale.
- Employee compensation, including benefits, must be comparable to compensation observed in other, similar enterprise groups; and all in the group should be fully, truthfully, and accurately informed about the bases for wage and salary administration so that the employees believe they are receiving fair treatment. Any study of ways to improve safety should be considered in the context of other desired objectives throughout all levels in the enterprise group.
- Absolute consistency in the application of rules is less important than considering each person individually. Some rules, such as wearing safety equipment, not smoking in an explosive area, must be rigidly enforced. Nevertheless, days off, overtime pay, improved equipment, and so forth, are individually important and may be individually considered. In practice, a member of a group almost always will choose being treated as an individual rather than being treated according to inflexible rules.
- Changing jobs several times in a career is becoming the norm. For many employees, this means transfers among work groups, but for some it has meant changing organizations. Managements have tried to lessen the shock of such change through broadened training, or out-placement programs, or

generous severance pay. Supportive efforts can help to sustain high performance levels, including a high level of concentration of safety factors.

Off-the-Job Safety Programs

Off-the-job safety is a worthy goal in its own right. It is good business. Analysis of lost-time off-the-job injuries and rates normally shows that they are more costly than lost-time on-the-job injuries. Encouraging employees to be safe away from work helps to preserve skills that could be temporarily or even permanently lost. It has the additional importance of keeping absenteeism low.

Further, being careful everywhere creates a good mindset. One's concentration on the job is likely to be enhanced by one's concentration off the job. And the plant or office is a healthier, safer environment because of it. The benefits also are widely shared throughout the plant or office communities.

Recommended Off-the-Job Safety Program

Especially for industrial units where an off-the-plant safety program does not exist, the following procedure may be helpful:

- Use a management letter, signed by the top manager, to announce the program to all employees. This letter should:
  —Explain that interest in off-the-job safety is legitimate. If an employee cannot get to work, the business is affected.
  —Request one-on-one contact between supervisors

OFF-THE-JOB INJURY REPORT	*Upon notification of an injury, the immediate supervisor (or the supervisor receiving the call) will complete this form. Immediately afterward, the supervisor will notify departmental supervision and Safety.*	
NAME OF INJURED	DEPARTMENT	AREA

DATE OF INJURY	TIME OF INJURY	IF VEHICLE ACCIDENT, WERE SEAT BELTS WORN?

NATURE OF INJURY

CAUSE OF INJURY

WHERE INJURY OCCURRED

IS INJURED HOSPITALIZED?	HOSPITAL NAME
HAS INJURED SEEN A DOCTOR?	DOCTOR'S NAME

CAN INJURED REPORT TO PLANT MEDICAL? ____ IF NOT, WHY________________________

CAN INJURED PERFORM LIGHT WORK? ____ IF NOT, WHY________________________

WHERE CAN INJURED BE REACHED BY TELEPHONE?

REPORTED BY	DATE	TIME

DISTRIBUTION: WHITE - DEPARTMENT SUPERINTENDENT YELLOW-SAFETY

Figure 3-7. Copy of sample report form for off-plant injuries.

and employees where management's purpose can be spelled out for each person.
—Set clear goals comparable to on-plant safety goals.
—While stressing the urgency of the program, *take a positive approach;* it should explain the benefits of good safety practices both on and off the plant.

- Establish an off-the-job safety committee with its own chair appointed from a high level of site management. Duties of this committee might include the following:
  —Set realistic goals.
  —Develop and administer well-detailed programs.
  —Compile and analyze injury data.
  —Analyze results.
- Establish a rapid reporting and communication system:
  —Distribute "What to do" cards to every employee.
  —Prepare an off-the-job injury report form to be used faithfully (Fig. 3-7).
  —Prepare a prompt depersonalized notice of the off-the-job injury.
- Combine short-term goals with awards and widespread recognition in order to establish peer pressure comparable to that achieved in on-plant safety programs.
- Maintain momentum:
  —Involve smart, energetic people.
  —Reward initiative and innovation.

Reporting an Off-the-Job Injury

Instruction of each employee is necessary. Distribution of and knowledge of the off-the-job injury report form may take care of this, but, just in case, each employee should be prepared to follow this procedure:

If you are injured and think you may lose time from work, call your supervisor as soon as conditions permit. If you are unable to call, have a member of your family call.

1. Give the following information:
   a. Date, time, and location of injury.
   b. Nature of injury.
   c. How injury occurred.
   d. Why it occurred.
   e. If an automobile collision, were there seat belts in the vehicle? How many? Were they being used?
   f. Has injured been treated by a medical doctor?
2. If the injured person has not seen a doctor but can move about, he or she should report to Plant Medical for examination. If a real lack of transportation will prevent the employee's coming to Plant Medical, Plant transportation will be provided to bring the employee in.
3. If the injured person has seen a doctor who has advised no work, not even light duty, department supervision will request Plant Medical to investigate. Work assignments sometimes can be rearranged for an employee if necessary to expedite his or her return to work.

The Importance of Maintaining the Record

An off-the-job safety program not only improves off-the-job safety performance but contributes to on-the-job safety. Employees should be encouraged to think of achieving and maintaining record performances on and off the job as part of the same plant and company goals.

The attitude of the employee work force is a major factor

in the success of a safety program. *It is important for everyone to understand why so much time and effort is directed at maintaining the safety record.* Call it team spirit, esprit de corps, momentum—that which arises from consciousness of and pride in being part of continued success must be sustained. Conversely, a break in the "success habit" that results from having an injury or failing to meet a goal can have a detrimental effect on a safety program.

It is not unusual for misunderstandings to arise, with attendant criticisms of supervision's attempts to stay within the organization's goals. For example, cynicism sometimes develops over special efforts to keep employees at work despite slight injuries, on or off the job. Management may provide special means to bring an employee to the plant and to provide light work assignments that differ from his or her usual duties in order to avoid a lost-time or recordable injury. However, the extra effort accorded each injury case is not intended simply to avoid recording the injury as such. It is to reap the long-term benefits that accrue from the success habit. The work force avoids a letdown, keeps its momentum, and shapes a positive attitude for achieving continued success. In doing so, the work unit may help avoid some other, more serious injury in the future.

This same philosophy applies to any and all safety efforts, but its application to the off-the-job program is particularly vulnerable to misunderstandings and criticisms of this type. A special management effort to get understanding and acceptance of this effort will have a beneficial effect on the program.

Going the Last Mile!

Some work units have found it helpful to encourage employees to report "near incidents," telling how an incident almost happened, and what was done, or could have been done, to avoid this danger. This event is publicized on bulletin boards—one employee's event, properly reported, may alert his or her coworkers to be on guard against a similar circumstance.

CONTRACTOR SAFETY

Every vendor must be knowledgeable about company policies and rules as they apply to vendors' activities, and be prepared to comply absolutely. Not every industrial customer has the same potential hazards, and not every customer is equally scrupulous.

It is especially important that vendors who deliver products at the plant site be monitored and be made to understand that carelessness in the handling of their vehicles and equipment, or failure to wear required protective clothing, eye shields and goggles, and safety shoes, may disqualify them as vendors.

For companies that have substantial business arrangements with contractors, it is helpful to prepare safety guidelines for in-house personnel to consult in making and maintaining those arrangements. Such a guideline might state that:

- Safety performance should be a prime consideration in the selection of all contractors.
- Contractors should not be used for jobs to avoid assignment to in-house employees unless the contractor personnel have special expertise and training.
- Contractors should be informed of all environmental, safety, and health hazards associated with company materials and processes being used. Communications spelling out this information should be documented.

- Safety- and health-related communications should be with contractor supervision and not directly with the contractor's employees, except when they are in immediate danger.
- Audits of *dedicated* contractor activities should determine compliance with contracts.
- Violations should be brought to the attention of the contractor's supervisor. If an unsafe act or condition creates an immediate danger of injury, immediate steps should be taken to stop the work.
- A log of contractor employee injuries and illnesses should be maintained.
- Contracts should be written so as to permit their termination upon a contractor's failure to comply with safety requirements.

Additional special cautions for contractors involved in new or revised processes are included in Chapter 4.

Transportation Safety

During the past two decades, an increasing number of headline stories have dealt with transportation disasters. Especially notorious events have included derailments caused by the deteriorating condition of U.S. railbeds. A number of incidents have raised threats of the release of toxic gases such as chlorine or the spillage of toxic or noxious materials into streams and rivers.

Hazardous materials are transported because they are needed. They are important to industry and the economic health of the country. Although the overwhelming bulk of these materials continue to be shipped safely, transportation incidents do happen. Even a seemingly minor mishap can become a major incident in the public eye and provoke a severe reaction against the shipper as well as the carrier.

Such incidents draw firefighters and police officers to the scene, as well as local, state, and federal officials who take an active interest in the handling of the incidents.

In the 1970s, such notoriety brought home to producers and shippers the need for manufacturers to take a proactive role in making sure that their products, especially hazardous materials, were shipped safely. Proper labeling had been accomplished several decades earlier, and a new design of tankcars had increased the integrity of shipping containers.

But failures continued to plague many communities situated on heavily traveled routes. Early in the seventies, corporate management at DuPont mandated much closer supervision to make sure that all regulatory requirements, including Coast Guard regulations, were scrupulously followed. Management also urged proactive efforts to improve the skills and attention needed for safer operations. Within DuPont both staff and operating organizations conducted in-depth studies, and the Safety and Fire Prevention Division was charged with monitoring activity across the company. The company's highly detailed Engineering Standards became the major communications vehicle. Additional cooperation with the International Maritime Organization was effected, as well as information exchanges with other countries, especially those in Europe.

The chemical industry in particular saw the need for establishing teams that could respond quickly to word of any disaster, and to be ready and knowledgeable about procedures to quarantine any dangers. In 1978 they developed the RHYTHM (Remember How You Treat Hazardous Materials) program.

Working with common carriers, shippers achieved major improvements in facilities, vehicles, and containers, and better standardization in labeling and documentation.

Under RHYTHM, companies get the right people to

the scene, making sure that they are equipped to limit hazards to the public and are adept at dealing with the news media, thereby minimizing the potential for harm to the company's reputation.

Ideally, the responsibility for informing the media rests with the carrier. In a railroad emergency, a "wreckmaster" or the conductor is the spokesperson. The operator becomes the spokesperson in a truck or a barge incident although the state or local police are the major source of information in highway incidents, and the Coast Guard or local police in river mishaps.

The expertise of these spokespersons may vary widely, so it behooves a company representative to guide and inform the carrier as to how hazards can be reduced and further damage to equipment can be minimized.

Although conditions today are far from perfect, the number of traffic incidents and equipment failures has been markedly reduced. The RHYTHM program today is more low-key than it was at its inception. However, the potential for deterioration of safeguards due to downsizing and streamlining, noted in connection with in-plant operations, is also a threat to transportation safety. Today, complaints are increasing about a "shortage of arms and legs" to do the job. Fewer people are going into the field. DuPont and other firms are finding it necessary to set up seminars to make sure that the necessary training of site coordinators gets done.

4 Process Safety Management

Process Safety and Risk Management Principles

To this point we have been concerned primarily with establishing safety programs that keep people deeply involved. Now we will turn to other, more technical aspects of safety.

Process safety and risk management may be defined as the application of systems and controls—programs, procedures, audits, and evaluations—to a manufacturing process in such a way that process hazards are identified, understood, and controlled, so that process-related injuries and incidents are prevented. Of increasing importance is the recognition that there are four important "stakeholders" involved with the safe handling of hazardous materials: the owner/operator, employees, the public, and government.

Figure 4-1. The photo shows experimental polymer film being extruded in a laboratory. Safety planning begins in the research laboratories and is part of the development of both potential commercial product and the equipment designed and fabricated to make the product.

Although the term "managing process safety" does not embrace all aspects of risk management, it is the heart and core of our responsibility as industrial managers. Speaking again in terms of our own experience, we would like to examine a DuPont program.

Process Design and Safety

From the very beginning of process design, safety must be an integral part of the design function. Every aspect of the process, ranging from the choice of materials to the disposition of wastes, must be examined with a view toward identifying and evaluating all risks and establishing in advance all contingency procedures for dealing with unplanned and unexpected emergencies.

Process Safety Management

DuPont's approach to process safety management was developed over many years, and the company has proactively assisted in the development of both professional and trade association guidance and regulatory requirements through the rulemaking process.

Management leadership and commitment form the foundation of any lasting effort to achieve and sustain excellence in process safety, which begins with the establishment of a safety culture as discussed in the preceding chapter.

Other management responsibilities important to process safety include:

- *Establishing process safety management policies and guidelines.* These are important in that they provide specific guidance regarding what needs to be done and how to do it.
- *Committing needed resources to implement process safety management policies and guidelines.*

Commitment of resources sufficient to implement process safety policies is essential in transforming a paper policy to an action program.

- *Involving employees.* Employees are uniquely knowledgeable about the process. Too often young managers are slow to recognize this. Often first line supervision and wage roll employees are the only "managers" immediately available. It is important that management provide for and encourage a broad spectrum of employee involvement in the design, implementation, and ongoing maintenance of process safety management systems.
- *Establishing clear accountability.* Performance should be measured against specific process safety goals and/or objectives. Accountability is a fundamental management principle, long established. People must not only understand what they are responsible for but also understand that they will be accountable for the results. Otherwise many things fall through the cracks, and management's expectations are not met.
- *Verifying performance.* Performance should be measured by degrees of compliance against established process management policies and guidelines. This verification is critical in the control and feedback loop. It provides an opportunity to acknowledge and recognize good performance, as well as to initiate corrective actions to address observed deficiencies.
- *Personal participation.* Personal participation in activities that visibly demonstrate process safety management commitment provides the opportunity for two-way communication with employees around safety, health, and environmental issues. This in turn gives employees the opportunity to see and understand:

—That management's conviction and commitment are deep and universal.
—That safety, health, and environmental issues are really "core values" and not just passing programs or fads.

- This understanding, in turn, helps an organization achieve the very desirable position of having *shared values,* a characteristic of organizations that have achieved operating excellence.

A Process Safety Management Model

The following analysis is highly technical and based specifically on the chemical processing industry where our experience has been concentrated. We believe, however, that the detailed analysis and procedural approach is broadly applicable across industry. The examples given demonstrate very well the complexity of safety procedures that are necessary in today's sophisticated industries and the remarkable efforts that have been made and are being made to ensure safety in a complex, technical environment.

Safety professionals from other industries, of course, must translate these methodologies into the efforts that are necessary when the hazards are very different from those applicable here.

Background[1]

An examination of the chemical and petroleum refining industry's performance provides some insights into the ex-

[1]The following material borrows liberally from a paper prepared and delivered by Arthur F. Burk, Senior Safety Fellow at DuPont, for a conference on Managing Safety and Health in the Workplace, conducted by the Institute for International Research in Chicago, IL, June 13–14, 1994. The title of the paper was "Achieving Excellence in Process Safety and Risk Management."

tensive and ongoing efforts needed to develop and publish guidance regarding process safety and risk management.

Accidental releases of hazardous substances, whenever they occur, present a potential for significant injury, environmental damage, and property damage. A brief look back provides some insights into the extent of that potential:

- An analysis of several data bases where injury and fatality data are provided indicates that 6158 fatalities occurred in 390 major process safety–related incidents over a 37-year period.
- An analysis of 170 of the largest property damage losses worldwide in the past 30 years by M&M Protection Consultants shows a total property damage loss of $7.35 billion dollars, or an average of $43 million dollars per loss.
- Further, a number of incidents have had some impact beyond the property line, affecting the surrounding community and environment. Some of these incidents are summarized in Fig. 4-2.

Incidents such as these have stimulated regulatory initiatives and trade and professional association guidance worldwide.

A Deluge of Guidance (1986–94)

Some of the incidents presented in Fig. 4-2 led to regulatory initiatives around the world. The cyclohexane explosion at Flixborough (1974) in the United Kingdom resulted in the CIMAH (Control of Industrial Major Accident Hazards) regulations in the UK in the early 1980s. The one- to two-kilogram release of dioxin at Seveso, Italy in 1976 led to the Seveso Directive, which

Background	Worldwide Industry Performance Loss/Fatalities
• 170 largest losses past 30 years*	$7,350,000,000 $43,200,000 (avg.)
Some Key Incidents	
• Flixborough (6/74)	28 fatalities $232,000,000
• Seveso (7/76)	Contaminated countryside
• Mexico City LPG (11/84)	300+ fatalities $20,040,000
• Bhopal (12/84)	2,800 fatalities 200,000 affected
• Chernobyl Nuclear Plant (4/86)	31 fatalities 300 mi^2 evacuated
• Piper Alpha Platform (7/88)	165 fatalities
• Pasadena, Texas (11/89)	23 fatalities $750,000,000
• Channelview, Texas (7/90)	17 fatalities loss–TBD

**Large Property Damage Losses* (14th Edition); M&M Protection Consultants; 1166 Avenue of the Americas, New York, NY 10036

Figure 4-2. Process safety incidents.

first passed in 1982. That directive requires all European Economic Community countries to adopt provisions necessary to prevent major "accidents," and it provides guidance for applying those provisions.

Such tragedies and the imperative to do something about them also had consequences in the United States. The LPG gas explosions at Mexico City (November 1984) and the methyl isocyanate release at Bhopal, India (December 1984) stimulated regulatory initiatives at both state and federal levels. In rapid succession the states of New Jersey, California, and Delaware passed laws and subsequently adopted regulations.

At the federal level, both the Occupational Safety and Health Administration and the Environmental Protection Agency began to prepare process safety management/ "accidental" release prevention–type regulations. The Clean Air Act amendments of 1990 provided the statutory authority. Section 304 authorized the Secretary of Labor to promulgate within 12 months regulations titled "Chemical Process Safety Management of Highly Hazardous Chemicals." Section 112(r) authorized the Administrator of the Environmental Protection Agency to promulgate Accidental Release Prevention Regulations by November 1993, which were to become effective three years later.

Concurrent with the regulatory initiatives described above, there has been an abundance of professional and trade association guidance. The American Institute of Chemical Engineers established the Center for Chemical Process Safety (CCPS) in 1985 and began publishing a series of guidelines. CCPS published its Guidelines for the Technical Management of Chemical Process Safety in 1989. The American Petroleum Institute published its Recommended Practice 750, entitled "Management of Process Hazards," in January 1990, and the Chemical Manufacturers Association began implementing the Process Safety Code of Management Practices as a part of Responsible Care, also in 1990. Compliance with Responsible Care is mandatory for membership within the Chemical Manufacturers Association.

Figure 4-3 illustrates the deluge of guidance that has occurred in the United States. For those wishing a comprehensive listing, Fig. 4-4 lists the different guidance documents in chronological order.

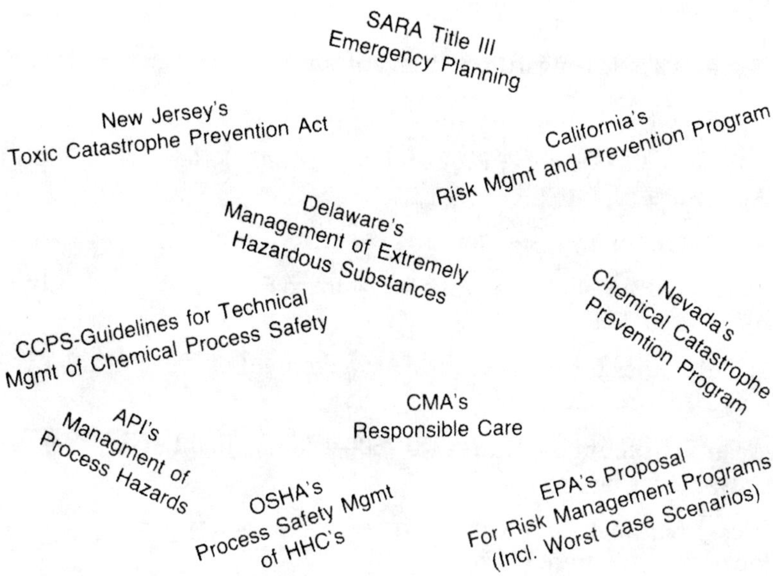

Figure 4-3. A deluge of guidance.

The Dilemma

The challenge facing most facilities handling hazardous substances is the question of what to do with all this guidance.

Or, to put it another way, how do people with responsibility for process safety management convert and integrate the deluge of guidance into a coherent, logical Process Safety and Risk Management Program (PS&RM)? A schematic suggesting priorities and/or sequencing is illustrated in Fig. 4-5.

From an operator's viewpoint, effective PS&RM is not ten programs bundled together with repeated instructions to "follow and comply with these programs." The ideal aid is one set of guidance that embraces all relevant regu-

GUIDANCE	DATE
New Jersey's Toxic Catastrophe Prevention Act	Jan. 1986
California's Risk Management & Prevention Program	Feb. 1986
Texas: Program to Detect & Prevent Catastrophic Releases of Air Toxics	Sept. 1986
SARA Title III, Emergency Planning & Notification	Oct. 1986
Delaware's Extremely Hazardous Substances Risk Management Act	July 1988
CCPS: Guidelines for the Technical Management of Chemical Process Safety	Nov. 1989
American Petroleum Institute—RP 750—Management of Process Hazards	Jan. 1990
Chemical Manufacturers Association—Process Safety Code of Management Practices	Oct. 1990
Clean Air Act Amendments of 1990	Nov. 1990
OSHA: Process Safety Management of Highly hazardous Chemicals	Feb. 1992
Nevada's Chemical Catastrophe Prevention Program	1992
EPA's Proposal for Risk Management Programs	Oct. 1993

Figure 4-4. Process safety and risk management guidance—chronological order.

latory requirements, including trade and professional association guidance, and is consistent with the company's philosophy and culture.

Four Key Steps

Development of a comprehensive and integrated Process Safety and Risk Management Program is best considered to include four key steps:

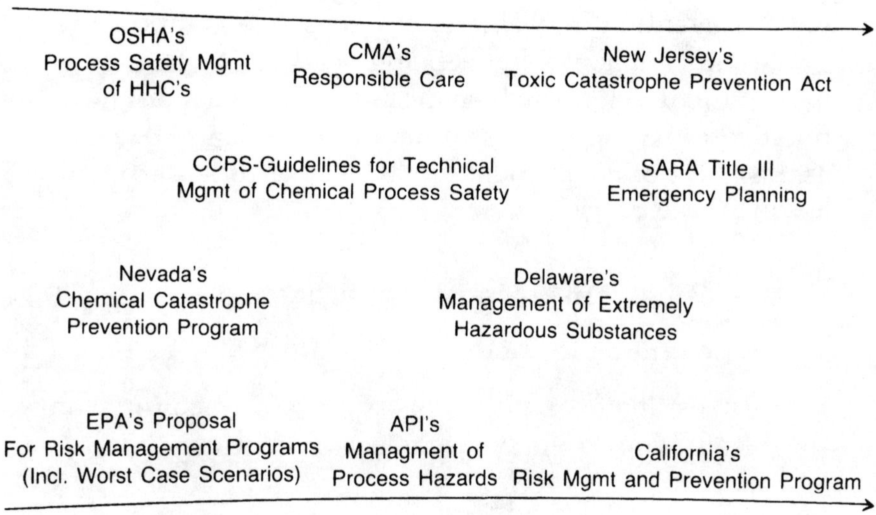

Figure 4-5. An integrated approach towards process safety and risk management.

1. Establishing the safety culture.
2. Securing management leadership and commitment.
3. Implementing a comprehensive and integrated Process Safety and Risk Management Program.
4. Achieving operating excellence.

The first two steps have been discussed at length in the preceding chapters; the third step is our focus in the following discussion.

Step Three: An Integrated Approach

Beginning with the Basics. A logical starting point is to look at federal guidance. Both OSHA's Process Safety Management Rule and EPA's Risk Management program have prevention and emergency response components. The significant new element being proposed by EPA is the requirement that facilities handling regulated substances must conduct hazard assessments (see Fig. 4-6). Hazard assessment, or consequence analysis, has at least the following three parts:

1. An estimate of potential release quantities.
2. A determination of downwind effects.
3. An estimate of potential exposures, including an evaluation of worst-case release scenarios.

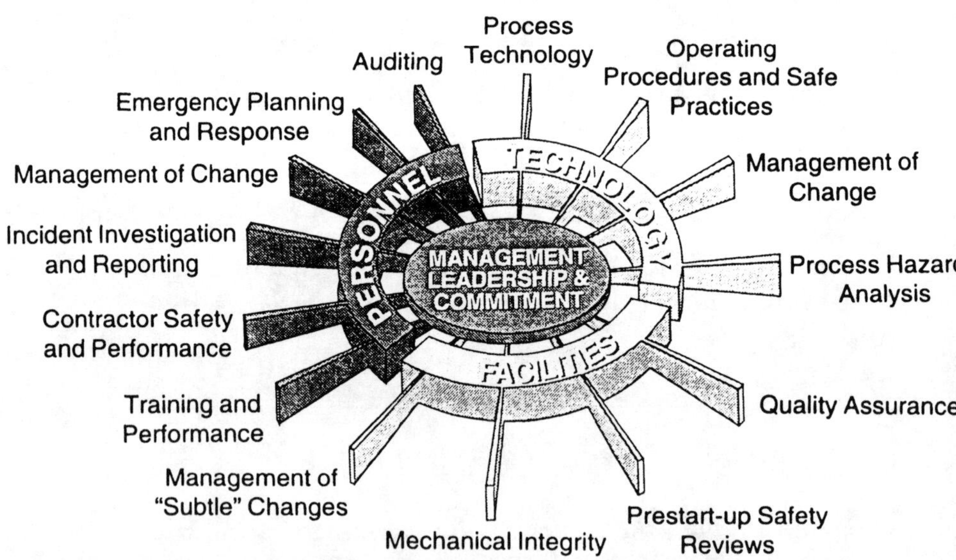

Figure 4-6. Process safety management model.

Process management should recognize and acknowledge the following important purposes of a hazard assessment:

- To provide an input into the process hazards review.
- To provide an input into emergency planning for the facility and the community.
- To provide a basis for risk communication with the public.

Because it is important that people responsible for the operation of a hazardous substance process have knowledge and understanding of the consequences of failure, it was decided to make consequence analysis an integral part of the process hazards analysis.

Actually, each process hazards analysis is to consist of two parts: a consequence analysis and a process hazards review. This approach embraces all of OSHA's regulatory requirements for process hazards analysis and *conceptually* embraces EPA's proposal for hazard assessments well in advance of regulatory requirements. Hazards assessments *based upon sound science,* with reasonable recommendations promptly carried out, benefit each of the four stakeholders—owner/operator, employees, the public, and government.

Getting into the Details

For comparative and illustrative purposes, the elements of a Process Safety and Risk Management Program have been organized into the following categories (see the model in Fig. 4-6):

- Management leadership and commitment
- Technology

- Facilities
- Personnel

Figure 4-7 summarizes the elements of DuPont's Process Safety and Risk Management Program and compares them with the corresponding OSHA/EPA elements.

The following descriptions of the 14 major aspects of the process safety management model follow a format designed for on-the-job reference. Consequently the 14 discussions frequently repeat elements that are common to different parts of the model. The 14 descriptions are subdivided into three sections, on technology, facilities, and personnel (see Fig. 4-6).

TECHNOLOGY

Process Technology

Principle. The process technology package provides a description of the chemical process of operation. It provides a foundation for achieving an "identification" and "understanding" of the hazards involved—the first steps in the process safety management effort. The process technology package generally consists of three parts:

- Hazards of materials
- Process design basis
- Equipment design basis

Essential Features:

- Hazards of materials—all pertinent physical and chemical property data for each chemical handled should be documented and communicated:
  —Boiling points.
  —Freezing points.

INTEGRATION:
"GETTING INTO THE DETAILS"

OSHA/EPA	*DUPONT'S APPROACH*
	MANAGEMENT LEADERSHIP & COMMITMENT
EMPLOYEE PARTICIPATION	INVOLVEMENT OF EMPLOYEES
MANAGEMENT SYSTEMS	SITE PSM COMMITTEES

TECHNOLOGY

PROCESS SAFETY INFORMATION*	PROCESS TECHNOLOGY
HAZARD ASSESSMENT PROCESS HAZARDS ANALYSIS*	PROCESS HAZARDS ANALYSIS CONSEQUENCE ANALYSIS PROCESS HAZARDS REVIEWS
OPERATING PROCEDURES* HOT WORK PERMIT	OPERATING PROCEDURES & SAFE PRACTICES
MANAGEMENT OF CHANGE*	MANAGEMENT OF CHANGE—TECHNOLOGY

PERSONNEL

TRAINING*	PERSONNEL TRAINING & PERFORMANCE
CONTRACTORS	CONTRACTOR SAFETY PERFORMANCE
	MANAGEMENT OF CHANGE—PERSONNEL
INCIDENT INVESTIGATION*	INCIDENT INVESTIGATION
EMERGENCY PLANNING & RESPONSE	EMERGENCY PLANNING & RESPONSE
COMPLIANCE AUDITS*	AUDITING

FACILITIES

	QUALITY ASSURANCE
MECHANICAL INTEGRITY*	MECHANICAL INGETRITY
	PRESTART-UP SAFETY REVIEWS
	MANAGEMENT OF "SUBTLE" CHANGES

Figure 4-7.

—Vapor pressure.
—Flammability limits.
—Thermal and chemical stability.
—Reactivity data.
—Exposure limits.
—Flash points.
—Ignition temperatures.
—Corrosivity data.
—Toxicity data.
—Hazardous effects of inadvertent mixing of different materials.

- Process design basis, documented and communicated:
  —Simplified process flow diagram.
  —Simplified process control diagram.
  —Description of the process chemistry.
  —Maximum inventory levels of hazardous substances.
  —Process steps and standard operating conditions (SOCs), including safety, health, and environmental consequences of deviation.
- Equipment design basis, documented and communicated:
  —OSHA requirements, including:
    - Materials of construction.
    - Process flow diagrams.
    - Protective safety systems.
    - Design codes employed.
    - Relief system design.
    - Electrical classification.
    - Piping and instrument diagrams.

  —Equipment files, archived and kept up to date.
  —Equipment design and installation, to comply with applicable laws and regulations, and with recognized and generally accepted engineering practices.

Operating Procedures and Safe Practices

Principle. The operating procedure provides a clear understanding of the detailed operating parameters and limits for safe operation, for those who are operating the process. It also clearly explains the consequences of operation outside process limits and describes steps to be taken to correct and/or avoid deviations.

Safe work practices include a carefully planned system of procedures and/or permits involving checks and authorizations before performance of nonroutine work in process areas.

Essential Features:

- Operating procedures must:
  - —Be consistent with the process technology.
  - —Include a comprehensive safety, health, and environmental section.
  - —Address start-up, normal operation, shutdown, and emergency operations.
  - —List standard operating conditions for each process step, indicating:
    - Maximum, preferred, and minimum conditions.
    - Consequences of safety, health, and environmental deviation.
    - Steps to correct and/or avoid deviation.
  - —Specify predetermined inventory limits.
  - —Describe safety systems and function.
  - —Describe instrument controls.
  - —Be kept up to date *at all times.*
  - —Be readily available to employees.
- Safe work practices shall be developed and written out. Employees shall be trained in their use to ensure the safe conduct of operating and

maintenance activities during nonroutine work. Examples include:
—Opening of process equipment and piping.
—Lockout/tagout of hazardous energy sources.
—Entry into confined spaces.
—Movement of heavy equipment relative to equipment containing hazardous substances.
—Integrity check of process equipment at time of turnover, but prior to acceptance.
—Entry and exit to and from a facility by support and contract personnel.
—Control of ignition sources (hot work permit).

Management of Change—Technology

Principle. Changes to documented process technology, including changes in materials used, equipment, and operating conditions, potentially invalidate prior hazards assessments. Accordingly, all changes to the documented process technology must be subjected to the same rigorous review that is applied to new processes.

Essential Features (of a rigorous review):

- Authorization document, which should include:
  —Written procedures to manage change of process technology. Such procedures should address:
    - Purpose of change.
    - Technical basis for change.
    - Safety, health, and environmental considerations.
    - Description of change versus current process technology.
    - Modifications to operating procedures.
    - Training and communication requirements.

- Limits for change—time and quantity.
- Approval and authorization requirements.

- Use of applicable controls: process hazard reviews, operating procedures, and piping and instrument diagrams.
- Closing document, which should include:
  —Summary of results and recommendations.
  —Document status of "open" process hazard review recommendations.
  —"Facesheet" controls listing modifications to operating procedures.
  —Tracking system to provide closure to both authorization and closing documents.

Process Hazards Analysis

Principle. Process hazards analyses are used to identify, evaluate, and control hazards associated with process facilities in a way that:

- Utilizes an organized, methodical approach.
- Seeks and achieves multidisciplinary consensus.
- Documents results for future use in follow-up and training of personnel—so that injuries and process-related incidents are prevented.

Essential Features:

- Consequence analysis, which includes:
  —An estimate of potential release quantity.
  —An estimate of downwind effects.
  —An estimate of the impact on surrounding populations, both workplace and off-site, and the environment.
  —A prototype plan for alerting people who may be

affected. A "dry run" should be made to ease a future alert and calm people if and when an emergency should occur.

- Process hazards review (existing facility), which should include:
  - —A review of the process using recognized methodology.
  - —A schedule per company guidelines.
  - —A multidisciplinary team with trained resource person/leader.
  - —Recommendations:
    - Communicated to personnel.
    - With a follow-up system to ensure resolution.
- Process hazards review (new facility), which includes:
  - —A screening review at the basic data stage, focusing on "conceptual issues."
  - —A preauthorization process hazards review, to include the scope of the work stage.
  - —A detailed process hazards review, to occur at the design stage subsequent to authorization of the facility.

Facilities

Quality Assurance

Principle. The quality assurance effort "bridges the gap" between design specifications and the initial installation. Quality assurance focuses on ensuring that process equipment is:

- Fabricated in accordance with design specifications.
- Delivered to the proper locations.
- Assembled and installed properly.

Essential Features:

- The design basis and criteria are documented and communicated to operating and maintenance personnel as part of the process technology package.
- The quality assurance program is developed, documented, and implemented to ensure that critical process equipment is fabricated and installed consistent with design specifications. The plan should include:
  —Written quality control procedures.
  —Appropriate checks and inspections during fabrication and constructions stages. It is imperative that people checking installation and fabrication be fully knowledgeable, be assigned full time, and have defined areas of accountability.

Prestart-up Safety Reviews

Principle. The Prestart-up safety review provides a final checkpoint to ensure that all appropriate elements of process safety management have been addressed satisfactorily.

Essential Features:

- Sites shall perform prestart-up safety reviews for all new and modified facilities. The purpose of the review is to ensure that:
  —Construction is done in accordance with design specifications.
  —The elements of process safety management have been appropriately addressed; that is:
    - Process technology.
    - Process hazards analysis.
    - Safety, operating, maintenance, and emergency procedures.

    - Training of operating and maintenance personnel.
    - Establishment of mechanical integrity systems.
    - Management of change documentation.
  - —The basic safety, health, and environmental considerations have been rechecked one last time.
  - —The facility can be started up safely.
- Prestart-up reviews should be conducted by a multidiscipline team.
- Reviews should be documented and signed by each member of the team.
- Follow-up systems should be established to ensure satisfactory resolution of recommendations prior to start-up with an approved outline of needed changes.

Mechanical Integrity

Principle. Mechanical integrity covers the life of the facility from the initial installation to dismantlement. Mechanical integrity focuses on system reliability to contain hazardous materials throughout the life of the facility. It addresses such topics as:

- Maintenance procedures.
- Training and performance of maintenance personnel.
- Quality control procedures.
- Equipment tests and inspections, including predictive and preventive maintenance.
- Reliability engineering.

The predictive and preventive maintenance topics are important and necessary to ensure reliable and incident-free operation. Such programs help prevent premature failure and help ensure operability of systems required for emergency control.

Essential Features:

- Maintenance procedures are established to ensure the mechanical integrity of process equipment on an ongoing basis.
- Maintenance personnel are trained in principles, functions, repair, and/or replacement of all parts of the mechanical system, including control systems and alarms.
- Quality control procedures are established to ensure that maintenance materials, spare parts, and equipment meet design specifications.
- A predictive and preventive maintenance program consisting of series of inspections and tests is established, with essential features to include:
  —Application listing of equipment and systems subject to tests and inspections.
  —Documentation of test objectives.
  —Documentation of test methods.
  —Inspection frequencies established.
  —Acceptable performance limits established.
  —Exception list issued for corrective action and follow-up.
  —Record-keeping system designed to facilitate review and analysis of test data.
- An ongoing reliability engineering analysis should be conducted for equipment critical to process safety.
- Site management should make appropriate use of consultants.

Management of "Subtle" Changes

Definition. Subtle changes are defined as any changes within the documented process technology that are not a replacement in kind. Examples might include rerouting a pipeline at a different elevation, which changes pressure drop considerations, or the use of valves from a different vendor that allegedly meet the original design specifications.

Principle. Subtle changes can lead to catastrophic events. All changes, including those that are within the documented process technology but are not replacement-in-kind, must receive appropriate review and authorization. The process safety management requirements for all such modifications must be established, and documented, prior to their authorization. The process safety management requirements must be completed prior to implementation of the modification.

Essential Features:

- Sites shall establish and implement written procedures to manage subtle changes.
- Site personnel must understand what constitutes change versus subtle change.
- Procedure for managing subtle changes should require:
  —Assessment of safety, health, and environmental considerations.
  —Documentation of process safety management requirements prior to implementation of the change.
  —Follow-up systems to ensure that process safety management requirements are completed prior to implementation.

Case Example. At one plant a vendor initiated "improvements" that caused a thermocouple to be ejected from a high pressure system, resulting in a major injury to an operator. Vendors must be required to notify users of *any* changes in equipment.

PERSONNEL

Personnel Training and Performance

Principle. Properly performing personnel are not only a key feature but an absolute requirement for the safe handling of hazardous materials. All other "key elements" of process safety management can be in place, but without personnel who are dedicated to consistently following documented policy and procedures, the chances of unsafe operation are high.

Having knowledgeable, well-trained employees alone will not ensure safe operations, free from human errors. The employees also must be physically able, mentally alert, and capable of using good judgment in following prescribed practices, and must have the will and the desire to do the job properly.

However, employees should be encouraged to alert management to changes they believe are desirable.

Essential Features:

(a) *Training*

- Site should develop, document, and implement training policies addressing:
  —Initial training.
  —Refresher training.
- Such policies should address these basic subject areas:

—Personnel requirements to provide and receive training.
—Qualifications of instructors.
—Basic skills training.
—Specific process or job task training.
—Emergency response and control.
—Refresher and supplemental training.
—Basic elements of effective training:
- Classroom training.
- Field training.
- Skill demonstration.
- Qualification testing.

—Record keeping.

(b) *Performance*

- Sites should develop and implement program to ensure fitness for duty.

Special Relationship between Training and Start-up. The start-up of a new plant or chemical process is a critical period fraught with problems, "glitches," frustrations, and—often—very expensive corrections and adjustments.

New product and/or new process development offers a special opportunity for building process safety from the ground up.

At DuPont, safety considerations start in the laboratory with process development. Basic data furnished to the design people will have properties of materials and equipment completely spelled out and toxic limits clearly defined. Recommendations are made for any special handling considered necessary.

People in the laboratory work closely with the design group as they do their work. At the same time, members of the plant groups who will operate the plant are brought in to review and study the design. Through safety reviews, based on an exploration of all possible contingencies (If this fails, then the following procedures must be instituted . . .), safety and emergency plans are updated with information that supplements the usual safety considerations of:

- Working space.
- Structural characteristics.
- Operability.
- Accepted design standards.
- Adequate safeguards to prevent:
  —Overpressures.
  —Overtemperatures.
  —Unsafe mixtures.
- Proper placement and design of relief valves and similar matters.

At the *start of construction*, the plant group must *start the training of operators, mechanics, and supervision*. This training proceeds simultaneously in several phases. Most successful plant start-ups have been accomplished as follows:

- Plant technical people are assigned to teach supervision from first line supervisors through successively higher echelons of management the process, chemical hazards, process control system, special equipment involved, and so on. These groups include both mechanical and production workers.
- Short quizzes at the end of each session are mandatory. Peer pressure greatly aids the learning process. Process safety considerations are thoroughly covered. Very early the fact that a *no injury* climate is not only desirable for humanitarian reasons but is good business for all is clearly enunciated and established.

  Management has been told by hard-bitten construction workers that DuPont is one company that recognizes "that people get more work done when they do not spend half their time making certain they stay alive." Wage roll groups thoroughly understand good business, and management's most creditable argument to them *is* good business.

But they also appreciate management's humanitarian interest, which must be shown in many small ways, such as rule flexibility in accommodating workers' personal needs.

- As training proceeds, members of mechanical groups including the instrument and electrical group, members of the production group, and members of the technical group are assigned in small working units with a designated leader of each to follow plant construction. This has been done effectively by dividing construction blueprints of a large plant into multiple segments. Each small-group leader marks each line, control system, and so forth. When construction is complete, the leader can be certain that it is correctly and safely done. One plant manager emphasized the critical need for doing the job correctly by promising to consider the discharge of a leader if as much as one gasket were missing on start-up.
- As supervisory training is completed, supervision of each group starts training the wage roll people in the group with the help and observation of technical engineers when needed. Quizzes to make certain every worker understands the process and safety considerations are mandatory.
- As all of the above proceeds, check sheets for each operator and many mechanical jobs are prepared by supervision, technical, and wage roll groups. These forms are put in a computer to be readily available

to the users. Computers and printers are available in shops and control rooms. It is important that each worker secure his or her own check sheet. Safety considerations are included in the check sheet.
- Also as the above work proceeds, safety and operating manuals are prepared, usually first by the operation and maintenance supervision, with technical personnel reviewing them in detail. The safety manual in particular is continually reviewed and discussed in group safety meetings.

This procedure as outlined has produced complex chemical plants that have reached full production in one week. Equally important, it has produced a coherent, high-morale work group thoroughly grounded in process and safety fundamentals and beliefs, with intergroup trust and respect.

Results achieved in untroubled start-up of chemical plants should be equally achievable with other kinds of production units.

Contractor Safety and Performance

Principle. All tasks must be completed safely in accordance with established procedures and/or safe work practices and must be consistent with the principles and the essential features of process safety management, whether the tasks are ultimately completed by site employees or by contract employees.

Essential Features:

(a) The site shall:
- Evaluate a contractor's safety performance and programs as part of the selection process.

- Establish clear lines of communication between the site contract administrator and the contractor.
- Inform the contractor of known potential fire, explosion, or toxic release hazards.
- Inform the contractor of applicable safety rules, procedures, and safe work practices of the facility.
- Explain the emergency response and control plan to the contractor.
- Maintain a contract employee injury and illness log.
- Periodically evaluate the contractor's performance in fulfilling his or her responsibilities.

(b) The contractor shall:

- Ensure that each contract employee has the necessary job skill training and is qualified to perform contracted work.
- Ensure that each contract employee is instructed in known fire, explosion, and/or toxic release hazards.
- Ensure that each contract employee receives and understands training regarding site safety rules and applicable safe work practices.
- Document the training provided, including means used to verify that employees understand the training.
- Ensure that contract employees follow safety rules and applicable safe work practices.
- Establish a program to ensure that contract personnel are "fit for duty" and are not compromised by external influences.
- Advise the site employer of any hazards found by the contractor or resulting from the contractor's work.

Incident Investigation and Reporting

Principle. Serious incidents and serious potential incidents will recur unless their causes are identified and corrected. Thorough and persistent investigation of all serious incidents and serious potential incidents will continually improve safety performance.

Essential Features:

- Sites should develop and implement an incident investigation procedure consistent with company guidelines.
- Initiation of the investigation must be prompt.
- Incident investigation reports should include and/or address:
  —The date and the time.
  —The date the investigation began.
  —A description of the incident.
  —Factual information.
  —A statement of the basic cause.
  —Identification of key factors, including equipment, human, and system deficiencies.
  —Identification of process safety management elements that need to be strengthened.
  —Recommendations for preventing incident recurrence.
- Incident reports should be reviewed with appropriate personnel.
- Incident reports should be retained for five years or longer.
- A follow-up system should be established to ensure resolution of recommendations.
- Relevant facts should be communicated to the Safety and Occupational Health (S&OH)

organization and to its safety counterparts at other appropriate locations.

Management of Change—Personnel

Principle. People are the one essential ingredient in all elements of process safety management. It is important to maintain a minimum level of (1) specific direct process experience and (2) knowledge and skill in managing process safety in the work force. Loss of minimum levels of experience and knowledge through personnel movements and organizational changes, such as changes in technology or facilities, potentially invalidates prior hazards assessments, which assume that knowledgeable people are present and in charge. Accordingly, personnel changes at all levels must satisfy preestablished criteria to ensure that minimum levels of experience and knowledge are maintained.

Essential Features:

- Criteria and guidelines must ensure minimum levels of knowledge and experience for both supervisory and nonsupervisory positions.
- Management must ensure adherence to those criteria and guidelines.
- Training must be provided for newly assigned personnel, including:
  —The principles and essential features of process safety management.
  —Technology and safety factors specific to operation.
- Line supervision must demonstrate proficiency within a predetermined time period.

- During weekly or monthly group safety meetings, safety and operating manuals should be reviewed.

Emergency Planning and Response

Principle. In-depth planning for potential emergencies is essential to ensure an effective response by site personnel working in close conjunction with supporting community emergency response organizations. The important products of these efforts are: (1) mitigation of the impact of the emergency on personnel, environment, and facilities; and (2) prompt control of the emergency situation.

Essential Features:

- Conduct a consequence analysis, and use the results as input for the plan.
- Develop a written emergency plan to provide for mitigation of the consequences. The plan should address:
  —Activation of the emergency response and control plan.
  —Notification of affected personnel.
  —Notification of emergency response organizations.
  —Notification of appropriate regulatory agencies.
  —Escape and evacuation routes and plans.
  —Accounting for all personnel.
  —Rescue operations including medical assistance.
  —Designation of primary and alternate emergency control centers.
- Develop a written emergency plan to terminate the release of hazardous material, to bring the emergency under control. The plan should address:
  —Emergency shutdown procedures.
  —Activation of emergency systems.

—Activation of the site's fire brigade or notification of the local fire department.
—Shutdown of adjacent facilities.
—Barricading of affected facilities.
—Activation of spill cleanup procedures.

- Train site personnel in implementation of procedures, including:
  —Periodic emergency *drills*—critical to the continuing improvement of employee understanding of emergency procedures.
  —Involvement of appropriate off-plant emergency response organizations.

Auditing

Principle. The only way to know how one "is really doing" is by making field observations and comparing performance with established standards. Proper auditing includes positive feedback on significant strengths as well as corrective feedback on areas needing improvement.

Essential Features:

- Sites are to develop and implement procedures for auditing process facilities for their compliance with site process safety management guidelines. Procedures should address:
  —Internal auditing; that is, auditing done by the line organization.
  —Outside independent auditing (site level), with:
  - Audits to be conducted using a prepared checklist for each element of process safety management.
  - Facilities to be audited at a frequency not to exceed once every three years.

—Results, which must be documented and reported to the site's safety and health committee.
—Audit recommendations, which must be promptly resolved.

- Other outside audits should be conducted by:
  —The business sector.
  —The corporate Safety and Health organization.

Achieving Business Excellence

There is a relationship between excellence in process safety management and the other facets that contribute to business excellence. Almost without exception, organizations that have achieved business excellence have achieved excellence in each of the activities that make up that business; that is, excellence in safety, excellence in quality, excellence in supply, and excellence in cost control. Further examination reveals a common thread in each of these supporting activities, and that common thread consists of:

- Sound, up-to-date technology and procedures.
- Trained personnel.
- Equipment that is maintained and reliable.
- Effective management of change.
- Auditing with control and feedback.
- A focus on doing each job the right way each time.
- High morale and group commitment, companywide.

Thus, the efforts to establish these common threads, which are an integral part of a Process Safety Management Program, are strongly aligned with efforts to achieve overall business excellence. Duke Power presented the results of a T&M in the redesign of their safety process at the

National Safety Congress in October 1994, and reported a marked reduction in total recordable injuries.

On the other hand, process safety management itself contains all the elements necessary for you to "run your railroad"!

5 Product Safety Management

Let us begin with some basic premises. DuPont's stated policy as a manufacturer of more than 1700 product lines is that it will produce only those products that can be:

- *Manufactured* safely, with no harm to its employees.
- *Transported* safely, with no danger to those who move these products on to customers along the way.
- *Used* with no danger by its customers and ultimate consumers.
- *Disposed of* without harm to the present population or to the populations in generations to come.

Documentation for this policy is contained in a 1915 Executive Committee memorandum.

This policy does not mean that any sector of the human population can live in a totally risk-free society. But a

Figure 5-1. Marketing programs of new applications for company products must undergo rigorous testing procedures to ascertain safety as well as efficacy. A major market for DuPont's Nomex fire-resistant nylon opened up when tests demonstrated that protective garments of Nomex provided superior protection for firefighters as well as industrial security personnel working in situations where fire constituted a possible hazard.

company can rigorously assume responsibility for identifying and controlling risks associated with its own products to eliminate injuries and ensure safety in their use.

Similarly, the Chemical Manufacturers Association, which is based in the United States but counts several internationally based companies among its 170 members, has launched a DuPont-supported program called Responsible Care. The association asks its member companies for a long-term commitment to continuously improve their performance in health, safety, and environmental quality. Compliance is a condition of membership.

More recently, in 1988, DuPont took a leadership role in the formation of the Council for Solid Waste Solutions. The council is part of the Society of the Plastics Industry, and its goal is to make plastics the most recycled material—by application—by the year 2000.

As forward-looking as such programs are, they cannot guarantee the absolute absence of risk. Every product sold, every service offered, carries with it potential safety, health, environmental, and liability risks. The point is, we must recognize and deal with risks to ensure safety, which is freedom from harm or danger.

In many cases, these risks are minimal; at the other extreme, some products may be highly toxic and require skilled handling under very close supervision to ensure their safe use. Most products tend to fall between these extremes.

A similar pattern can be discerned in the service occupations, which often are involved in the distribution, handling, and use of products supplied by other vendors. In addition to the inherent risks of some materials, there is also a factor of added risk in the mishandling or the misuse of the products.

Every mangement team must analyze, define, and provide instructional materials pertinent to dealing with all risks that are part of its business operations. In individual firms, the most effective way to manage these risks is with a strong product safety management program.

In practice, many products are closely regulated at the federal or the state level in their manufacture, processing, storage, transportation, or marketing. At DuPont, for example, these products have been part of such product lines as explosives, heavy chemicals, ammunition, crop protection chemicals, pharmaceuticals, medical devices, and radioactive materials. Over the years, government agencies working with industry have developed extensive guidelines that can be incorporated into an individual firm's safety procedures. The ultimate responsibility, however, remains with managment.

Safety Management Systems

A properly implemented product safety management system will:

- Demonstrate to the public, government, and customers that the company intends to be a socially responsible supplier of products and services.
- Assure compliance with applicable safety, health, and environmental laws and regulations, as a minimum, and help avoid needless "product bans" and excessive regulation.
- Make product safety management a marketable feature of the company's products and services.
- Promote product safety management as another element of the customers' safety, health, and environmental protection efforts.
- Assure that intermediate custodians (for example, warehouse, terminal, and service center operators; distributors; and carriers) understand and accept the producer's safety policies and commitment to product and service safety.

It is in the producer's self-interest to work carefully and cooperatively with those who are in the producer-to-consumer chain. In an atmosphere of good faith, all firms involved should be able to agree that the maximum effort to adopt safety procedures at every stage is in everyone's best interest. All organizations involved should:

- Manage product safety as an integral responsibility of the company's business center.

- Expect and reward excellence in product and service safety.

Product safety management is properly a line organization responsibility. As such, it must be aligned with profit responsibility. At the same time, the safety oversight usually is best vested in a supportive staff structure.

Product safety management focuses on the prevention of adverse effects on the public and the environment by the introduction of a particular product or service. This responsibility extends from the development of the product, through the manufacturing process, where it dovetails with the broad safety program of the plant, and through marketing, transportation, distribution, and disposal.

The product safety management system ensures that, where appropriate, the following activities are carried out:

- A customer's use and disposal of a product are evaluated.
- Training tools and communications including labels and material safety data sheets (MSDS) are prepared, kept up to date, and distributed.
- Interaction with governmental agencies, which may be required to regulate the product or service, occurs in a timely manner.
- Applicable environmental guidelines and other standards are addressed.
- Product safety records are maintained.

Oversight at the Top

In a large, multinational organization, corporate management may have worldwide responsibility, but, by mutual agreement, corporate management may delegate assurance of compliance with foreign laws and regulations to a designated individual or office in the foreign region. Such delegation would include any interaction with non-U.S. governmental agencies—for example, those in Europe, Asia-Pacific, and Latin America.

Product Evaluation and Classification

Each existing product and service and all new products and services should be evaluated to determine the basic risks the product or service presents. Some of the risk analysis or assessment may draw upon risk assessments completed at an earlier stage in process safety management, as discussed in Chapter 4.

A firm with many and varied product lines probably will find it helpful to classify its products or services according to their potential harm. Such a classification will assist product handlers in prioritizing these risks to make sure they are properly addressed.

Such classifications will vary from firm to firm according to the nature of the products they offer. But the following categories established at DuPont may serve as useful examples:

High risk:

- Highly or extremely toxic materials.
- Explosives, ammunition.
- Radioactive materials.
- Etiological agents.
- Pharmaceuticals, medical diagnostics.

Moderate risk:

- Products of combustion hazards.
- Crop protection chemicals.
- Most services.

Low risk:

- Firearms.
- Instruments, machinery.
- Most fibers, polymer films, and resins.

Toxicity Data Base

In some cases the required toxicity data base will be dictated by regulatory requirements—for example, those products under the oversight of the Federal Insecticide Fungicide and Rodenticide Act (FIFRA), the Toxic Substances Control Act (TSCA), and the Food and Drug Administration (FDA). These requirements will vary from one jurisdiction to another.

In any case, the toxicity data base has to be adequate so that the necessary information is available for users to assess and properly manage the potential harm presented by the product in question. The adequacy of the data will depend on several factors, including intended use of the product, user experience, potential exposure, and number of people potentially exposed.

Especially where high and moderate risk categories are involved, toxicity testing is dictated by regulations. At DuPont new product development falling outside government regulatory guidelines requires consultation at an early stage with the company's Haskell Laboratory, described below.

For companies whose operations cannot justify a major in-house toxicity testing program, trade associations such as the Chemical Manufacturers Association may provide needed services.

Product Safety Management Reviews

For the first five years after the introduction of a new product, the learning curve tends to be very steep, and profit centers should consider frequent—at least annual—reviews. After that introductory period, regular though less frequent reviews are advised, with their frequency dictated by the level of assessed risk.

Other factors that should trigger a product review may include:

- A significant new use.
- New toxicity data.
- A serious incident.
- A significant increase in volume.
- A licensing or joint venture agreement.

The product safety management review should include occupational health and environmental safety factors.

Product safety review participants should be chosen from appropriate support groups and represent a wide range of related disciplines with broad knowledge and skills, such as the customer service, legal, environmental, occupational health, process safety, fire protection, toxicology, distribution, warehouse management, marketing, public affairs, engineering, and production functions.

Toxicity and Toxicology

Although some element or degree of toxicity is a potential safety problem in every industry, this is a major focus of attention in the chemical and allied industries. Therefore, we will take a special look at toxicity, the science of toxicology, and DuPont's own program for dealing with the safety issues of toxicity.

The science of toxicology is almost as old as manufacturing, but it has really come of age in the past two decades. This growth is a tribute to the hundreds of dedicated scholars and administrators in academia and government as well as industrial research organizations, who have helped to expand the frontiers of knowledge in the discipline.

However, many more chemicals are present or being made available in the human environment than can be evaluated for potential toxicity with available methods and resources. Estimates based on the Chemical Abstracts Services (CAS) Registry in the United States indicate that the universe of chemicals consists of more than five million known entities.

In 1980 the United States National Toxicology Program perceived an urgent need to get a better fix on the magnitude of this problem. In conjunction with the National Institute of Environmental Health Sciences, the program contracted with the National Research Council and the National Academy of Sciences for a study, with two principal charges:

1. To characterize toxicity-testing needs for substances to which there is known or anticipated human exposure. Such information is critical if federal agencies responsible for the protection of public health are to be able to assess the toxicity of each substance.
2. To validate uniform and wide-ranging criteria by which authorities can set priorities for research on substances with a potentially adverse public health impact.

Four principal committees worked over the next four years to meet this challenge. A "select universe" of 65,725 substances of possible concern was identified.

It was manifestly impossible to study more than a tiny fraction of that large universe. Through a random sampling process, 575 substances covering seven major end-use categories were selected. From this sample, a subsample of 100 substances was selected by screening for the presence of at least some toxicity information. In-depth examination of this subsample led to the conclusion that there was enough toxicity and exposure information for a complete health-hazard assessment for only a small fraction of the subsample. By inference, similar conclusions were made for the select universe of more than 65,000 substances.

The full range of findings of this massive report is much beyond the scope of this chapter, but a couple of observations are relevant here:

- Industry must expand greatly both the skills and the scope of toxicology if it is to protect populations in the United States and abroad.
- In the absence of adequate funds to meet all needs, industry also must establish priorities that can draw support from a worldwide consensus.

DuPont's Haskell Laboratory[1]

Haskell Laboratory was founded more than 50 years ago. It was the first industrial toxicology laboratory in the United States and one of the first in the world. It remains one of the few full-service industrial toxicology facilities in the United States.

Haskell is an integral part of DuPont's program in safety, health, and environmental protection. The laboratory currently has a staff of more than 230 people (see, e.g., Figs. 5-2 and 5-3). About 20 percent of these employees have doctorates in medicine or veterinary medicine, as well as specialties in such disciplines as molecular biology and microbiology, aquatic toxicology, analytical chemis-

[1]Most of the material describing the role of the Haskell Laboratory in the expanding science of toxicology is derived from a speech delievered by Dr. Charles F. Reinhardt at an international forum in Bangkok in July 1987.

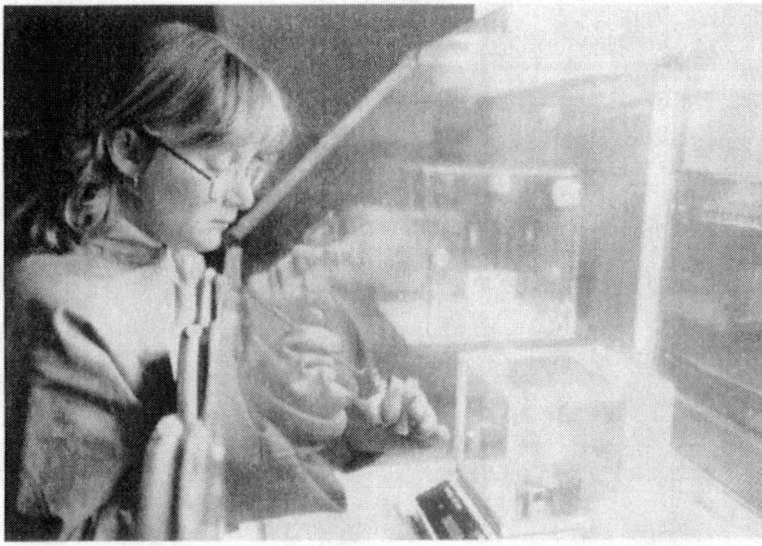

Figure 5-2. Safety of products: A biologist at DuPont's laboratory for toxicology and industrial medicine studies potential hazards of new materials under investigation to assure their safe manufacture and use.

try, biochemical toxicology, developmental toxicology, and pathology.

Haskell Laboratory's function is to assess the toxic properties of chemical compounds that DuPont develops or uses and to determine the precautions that are necessary to protect people and the environment. Through testing, consultation, and research activities, the laboratory determines the degree of toxicity and acceptable exposure levels at which a material can be manufactured, used, and/or distributed safely.

Historically, the primary intent of industrial toxicology programs was to determine the toxic properties of substances so that employees were protected. Today, DuPont's program, as developed in response to the needs of the company and the broader society, considers total product

Figure 5-3. A biologist at the Haskell Laboratory examines mouse lymphoma cells in a short-term test for possibly carcinogenic chemicals. Only cells that are mutated by the chemical are visible.

safety. Product safety evaluation is a significant focus, from discovery and development of a compound to disposal of any wastes.

DuPont's present toxicology program has five principal objectives:

1. To provide toxicity data on chemicals made and used in manufacturing processes.
2. To identify, prior to production, any chemicals that may be toxic so that safe handling procedures can be instituted.
3. To provide information for industrial hygienists and occupational physicians.
4. To provide a technical basis for establishing safe exposure limits.
5. To provide technical information required by government agencies.

Studies of some compounds may take several years to complete. The findings of some of these studies may be critical in deciding whether DuPont will produce certain substances at all and, if it does, how the compounds should be manufactured. Also, many of the studies are required for the company to comply with laws and regulations of governments around the world.

Toxicologists begin with an axiom of their profession: virtually every substance is potentially toxic. As Paracelsus observed more than 400 years ago: "Dosage alone determines poisoning." By varying the amount and the frequency of exposure, it is possible to identify a no-observed-effect level for the conditions studied. This information then can be used to set an acceptable exposure level at which a hazardous substance poses no significant risk.

The classic procedure, observed at Haskell, is to use laboratory animals—primarily mice and rats—as models to determine effect and no-effect levels, and to project those findings to human exposure. Haskell scientists are not ready yet to subscribe to the recent Bruce Ames assertion that "using rats to predict the risk of cancer is a 'bankrupt' approach [that is, one that may produce more false leads than helpful ones]," but toxicologists do listen carefully to what Ames is telling his fellow scientists about the dangers of extrapolation.

Testing covers both acute and chronic studies as well as routes of administration such as inhalation, ingestion, skin absorption, and eye contact. Measurement in parts per billion is not uncommon. With very sensitive analytical instruments, scientists now are sometimes discovering contaminants that they did not even suspect existed.

A variety of research technologies, including computer models and bacterial and cell culture systems, are being used to detect potential health problems or undesirable side effects (Fig. 5-3). However, these techniques do not fully represent the complexities of the human system. Therefore, laboratory animals are used to help predict human response. Company policy is to use laboratory animals only when necessary, and then in the most humane way possible.

In addition to the program outlined above, studies may be done to determine the metabolism of the tested chemical along with its disposition in the test animals (absorption, excretion, retention, and biotransformation). Also studies may be carried out to determine the mechanism of action—that is, how the chemical produces its effect. These studies often provide very good information for carrying out a scientifically based risk assessment.

In the following paragraphs we suggest several strategies that toxicologists might consider, and, in a few cases, offer examples from DuPont's programs.

1. *Toxicologists can promote better sharing of research among the community of scientists.* As a corollary of this, toxicologists must make sure that the results of this research flow through the industrial safety pipelines so that people in product safety management have ready access to the findings.

At present, the U.S. Environmental Protection Agency

(EPA) has been working with similar organizations in other countries, particularly the Organisation for Economic Cooperation and Development (OECD), to find ways to get better data on existing chemicals. In cases where preliminary testing of chemicals indicates need for further study, additional testing of those chemicals would receive high priority to resolve remaining questions.

2. *Toxicologists can establish improved protocols for the evaluation of new chemicals coming onstream as well as for many of the older chemicals that have been taken for granted.*

3. *Toxicologists can focus more on the really bad actors and show less concern about the materials that appear to be innocuous.*

Toxicologists have the best insights into what materials are truly threatening, and this list is not necessarily congruent with the chemicals sensationalized by the news media. Producers of a number of viable food products—in the United States, cranberries and apples are examples—suffered severe financial losses as a result of media-engendered hysteria.

Whatever the outcome of such reexamination, industrial toxicologists will be forced by the constraints of limited resources to make early decisions on what substances need concentrated attention and what studies can be deferred in the absence of conclusive evidence.

4. *Better communication can be practiced with the regulatory agencies, the news media, and eventually the general public to prevent priorities from getting skewed by uninformed or misinformed activists.*

New and extraordinary demands have been made on industry scientists to publish vast data on what toxic materials they work with and what volumes of toxic end-products or by-products "are released" to the broad environment.

No doubt the greater availability of some of these data will be of some use to public agencies that have ongoing responsibilities for public health and safety. However, media treatment of these data often has been less than helpful. Reporters, and their editors in some cases, have unjustifiably made members of the chemical industry look like villains.

The volumes of the chemicals involved may have little relevance to actual hazards. Many of the chemicals targeted in these revelations are being disposed of by methods toxicologists believe are thoroughly safe. In most cases, these chemicals do not cause and have not caused problems for anyone. These disposal methods are not to be equated with the toxic waste dumps of previous eras, which were often of great significance but are gradually being neutralized.

Toxicologists can play a critical role in realistically evaluating these issues.

5. *Toxicologists can give risk assessment a more constructive role.* As a corollary, those who work in the industrial sector must do a better job of emphasizing the great efforts being made at great expense to make manufacturing plants and processes safe.

Industry managements do better when they share the data of risk management with those agencies that need to know, while developing programs to make *all* procedures as safe as is technically possible. At the same time, industry spokespersons need to develop programs to make this commitment to improved safety evident to local media and plant neighborhoods. This is primarily a management responsibility, but industrial toxicologists can assist management in giving its immediate publics a better understanding of how small these risks may be in an industrial world where nothing is ever totally risk-free.

6. *Toxicologists can provide significant inputs in the innovation of new products.* Early involvement of toxicologists in product development can be important in helping chemists to synthesize products that are efficacious and low in

toxicity. Also, potential toxicity problems can be detected early and appropriate precautions taken.

Toxicologists have been able, through the use of mutagenicity assays, to test chemical analogues and allow chemists to modify them to eliminate mutagenic activity while retaining the desired efficacy. Thus, researchers can develop safer products while reducing the number of chemicals that otherwise might undergo extensive testing before the undesired effects were discovered.

Let us mention one product development effort of our former company. Over the past decade DuPont has developed a number of products based on sulfonyl urea chemistry, significantly aided by toxicological research. These products are essentially preemergence herbicides that can eliminate certain noxious weeds through the application of very small quantities of chemicals. *Properly applied,* three or four thimblefuls of these herbicides can successfully treat as much as three acres of tillable land. In these minute applications, the chemicals quickly biodegrade and pose no hazard to agricultural workers.

Environmental Protection

Up to this point we have emphasized the role of Haskell Laboratory in assuring human safety. Let us turn now to a brief account of DuPont activities in support of environmental protection.

Any potential impact of DuPont activity on the aquatic environment is likely to become a priority of Haskell's aquatic toxicologists. Their tests may cover everything from complex industrial effluents to products that may be marketed.

Although the aquatic group tailors specific testing protocols to meet international requirements, the most basic study involves static toxicity testing in which aquatic species, generally fish, are exposed to test compounds in standing water to determine lethal concentrations or to produce a specific effect. Dynamic acute studies, on the other hand, involve using constantly renewed exposure solutions in flowing water, a test that is more relevant to the real environment.

Daphnia or fresh water fleas are used in chronic studies to determine the effects of compounds on reproduction and growth as well as short-term lethality. DuPont also tests the ability of fish and shellfish to absorb a compound such as a heavy metal dissolved in water, to determine whether it can be passed along the food chain.

Resulting information from most, if not all, of the tests is likely to be published or shared with government agencies, which recommend and administer myriad laws and regulations on the environment, public and employee safety, and health. The company also has regular contacts with the academic and scientific communities. Indeed, almost everything Haskell Laboratory does is intended for use outside its walls.

These activities relate, of course, to concerns of the chemical industry and its desire to avoid harm to employees, customers, and the environment. Industrial managers are increasingly finding that they have cradle-to-grave responsibility for materials that they generate, including finished products and wastes. But this considered approach is also good business.

Early involvement of toxicologists enables DuPont to identify potential toxicity problems before product development goes too far, and to institute protective measures early. And it is just possible that early identification of a toxicity problem, before significant development expenses have been incurred, may give research and business organizations a new direction for a successful and safe product.

In conclusion, the expanding universe of needs and the

efforts made to supply those needs place severe pressure on toxicologists everywhere. The broad objective is to make sure that public safety is adequately considered.

However, the resources of any one industrial firm are severely limited. The strategies suggested above should help everyone to make the most effective use of those resources. The strategies are not at all revolutionary; to some extent, they are already in place.

For industrial toxicologists, the challenge is to use their developing knowledge and skills to define the many parameters of safety.

For those who are administrators the challenge is:

- To make prudent decisions in choosing research programs.
- To assign the skills of research teams according to well-thought-out priorities.
- To make sure that discoveries are understood both in the halls of government and in the management suites of business organizations.

Quite literally, the safety of hundreds of millions of people is in the hands of industrial and academic toxicologists.

Special Problems

Despite the admirable advances in toxicology, many problems have come to the fore in the past 20 years, and many more undoubtedly will have to be addressed with new experience. The response today, however, is swift, and the safety screen for passing or holding up new products is highly effective.

Long "Incubation" Periods

We noted in Chapter 1 that materials that have been produced for decades seemingly without deleterious effects can suddenly appear to be causative factors in slowly developing health problems. We offer two case examples with which DuPont has had direct experience.

Motor Fuel Additives. Tetraethyl lead (TEL) and tetramethyl lead were developed by General Motors more than 60 years ago as anti-knock agents, critical as gasoline additives for use in high compression engines. Minute amounts of these additives greatly improved performance; billions of gallons of such motor fuel have been blended, distributed, pumped into automobiles, and used without any direct evidence of harmful effects.

The toxicity of lead and lead compounds has long been recognized, and the manufacture of the lead-based additives required highly skilled and monitored processes. DuPont took on this chore, and has been the largest producer of these materials worldwide. It was believed that the products of combustion in internal combustion engines dispersed whatever solid lead particles remained in such a diffused form that no toxic ingestion was likely.

In recent years, however, public concern about traces of lead poisoning suspected in children in urban communities—perhaps induced from a variety of sources including old-formula lead-based paints—has led to the determination to proscribe the use of all lead-based compounds whenever possible.

Now new technologies for achieving high octane gasoline have reduced public anxiety about "lead poisoning" although evidence linking microscopic fallout of lead particles from gasoline consumption to mental retardation is ephemeral at best.

Aerosols, Refrigerants, Halogenated Cleaning Compounds. About the same time that the scientists at General Motors were developing effective anti-knock compounds, their scientific teams were finding solutions to other pressing problems resulting from the toxicity and/or the flammability of compounds used in the exciting new technologies of refrigeration and air conditioning. They found that a series of relatively inert chlorofluorocarbons could provide highly efficient cooling agents, which could be used with almost complete safety in both industrial and consumer applications.

Again, the manufacturing processes were potentially hazardous, and DuPont was asked to take them on. The refrigerants became standard worldwide, and other applications of these chemicals—especially in aerosols—provided rather inexpensive and safe dispersing agents. This class of chemical compounds seemed foolproof: an industrial contribution of enormous convenience for everyone with no evident drawbacks.

Thus it came as a major shock when academic scientists began to throw up red flags. Researchers had set up computer models attempting to explain newly detected variations in the seasonal cycles of thickening and thinning of the ozone layer in the stratosphere. The scientists posited a slowly accelerating degradation of the ozone layer with increased exposure of plant and animal life to ultraviolet radiation, already suspected to be a contributing factor in the most malignant form of skin cancer.

And the scientists identified aerosols, particularly the lighter chlorofluorocarbon compounds, as the triggering agent in the destruction of atmospheric ozone. Worse yet, the compounds, relatively inert in the life zones close to the earth's surface, broke down in the high level exposure to sunlight, setting up chain reactions that would accelerate the destruction of ozone.

Although the changes had not reached dangerous levels, the scientific models predicted an intensification of ultraviolet radiation in the decades ahead. So the scientists called for a rapid decrease in the production and an ultimate cessation in the use of the chlorofluorocarbons.

Such a changeover would require retrofitting of many industries at costs of billions of dollars worldwide. Although the empirical evidence is still not wholly conclusive, manufacturers of the cooling–freezing compounds are accomplishing the necessary changeover. The safety of plant and animal life is paramount.

Customer Misuse

Whether willfully or through ignorance, customers around the world daily suffer many injuries through the misuse of products that are highly effective when used properly. The education of the customer, especially the ultimate consumer, is an ongoing, monumental task. We offer two example from DuPont experience.

Crop Protection Products. Many herbicides, fungicides, and pesticides work miracles in sustaining profitable production of high quality agricultural products. These crop protection chemicals, however, are usually complex chemical compounds that function because of carefully measured levels of toxicity, destroying what needs to be destroyed—weeds, pests, fungi, and other plant diseases—while preserving what is valuable to useful botanicals and useful animal life, including humans.

Failures in following instructions, so that too much product is applied in the wrong places, or applied by workers not using recommended clothing, face masks, etc., can lead to injury or even death. Crops can be destroyed or rendered unmarketable. Lawns and shrubs can be blighted. Wildlife, too, can be killed or injured. Also, some marginal farmers or migrant workers are not fully literate; some families, for example, have been made seriously ill by consuming disinfected seeds.

Packaging Materials. A few decades ago, a raft of fatal or near fatal injuries resulted from secondary uses of transparent plastic films, sometimes as playtime shields by uninstructed children. Today, garment bags and similar applications carry better warnings and ventilating holes that practically rule out asphyxiation.

Industrial managements have not only a social responsibility but a self-interest in making sure that their products are properly labeled with necessary instructions and warnings boldly visible. And they should cooperate with other businesses in the maker-to-consumer chain to make sure that necessary information is carried along at every step of the marketing process.

Changes in Product Liability

Once producers and distributors were protected from liability in cases where the persons damaged were negligent, willfully or otherwise, in their use of a product. That protection has largely evaporated. Especially in jury trials, damages often must be compensated by the party with the deepest pockets—all the more reason for going several extra miles in discouraging misuse of products.

In April 1994, DuPont and several other suppliers announced their withdrawal from those markets for plastics and polymers used in making permanent medical implants. These implants include heart valves, pacemakers, shunts, replacement blood vessels, and artificial joints.

These devices often are surgically installed to keep patients alive, to relieve a dangerous or a limiting condition, or to replace a worn-out natural part, such as a knee or a hip joint.

In many cases, only a small amount of the polymer material is used for implants. A resin used in vascular grafts and surgical patches, for instance, also is widely used in general industry.

Clearly, these products have been a major medical advance, and the materials themselves, properly employed, can be used safely. However, in recent years the product liability laws have allowed plaintiffs to sue any company involved in any aspect of a product's manufacture or use.

As one example, DuPont noted that it had been sued 258 times over a single product that made use of a small amount of its Teflon tetrafluorethylene material. Whatever failure may have been alleged, such failure was not due to the quality of the Teflon material. The company's legal fees had reached about $8 million a year. In such cases, product safety is a necessary prerequisite, but it may not be sufficient for avoiding liability in commercial applications in all cases.

Turnover in Personnel

The human population is a passing parade. We noted the extreme importance of taking this factor into account in hazard assessment (discussed in Chapter 4). But this phenomenon is true in the marketplace as well as in the factory. As people come and go, educational processes get dangerously watered down. So the education of the customer, including the consumer, must be endlessly repeated. No matter how successful, no safety program can ever rest on its laurels.

Radical New Technologies

Technology, too, is a passing parade. Major improvements may call for major new products as well as major new processes. Sometimes these developments push operations outside the bounds of present knowledge and require vigilant surveillance of all aspects of safety, with a readiness to modify process design, materials, and employee work procedures as new information becomes available. A classic example of this is the nuclear energy activity ushered in during World War II. We give detailed reviews of DuPont's experience in this field in Chapter 9.

Figure 6-1. Concern about safety outside the fence, including the health of the plant environment, is demonstrated by the streams studies conducted by noted biologist Ruth Patrick. These studies were conducted near many of DuPont's plants to make sure that no significant concentrations of hazardous materials were migrating to local ecological systems.

6

Industrial Safety and the Plant Community

We in industry are concerned not only with the prevention of incidents on the plant or at the office or the distribution center but also with the mitigation of risk to the community at large. We also care about community perceptions of risk arising from uninformed rumors and vague fears generated by the vast amount of catastrophic news in the media. Any discussion of this phase of risk management must be prefaced by several premises, which we present as a basis for further discussion.

1. Risk is a fact of life. It has been a factor in human endeavor for centuries and will remain so for centuries to come. The late renowned scientist René Dubos once observed that

> there is no reason to believe that the dangers to health arising from modern technology are greater than the

> dangers that arose in the past from the natural forces to which man was exposed and which he has overcome in the course of his long social and biological experience.... Now, however, most changes are explosive, spreading over a whole continent within a few years, and affecting everybody almost simultaneously.
>
> To demand a certified verdict of safety before accepting a new technological innovation would clearly result in paralysis of economic and technological progress.... A society that does not continue to grow through adventure and willingness to take chances is not likely to survive long in the modern world. With regard to health as to all other fields, society must be willing to take educated and calculated risks inherent in a technological civilization.[1]

2. When we talk about risk management we are not talking about the elimination of risk altogether, which is impossible. We mean *keeping risk within manageable limits.*

3. The prevalence of risk is multiplied and compounded somewhat in proportion to the growth and application of technology. At the same time, the growth of technology gives us better means of controlling risk if we are wise enough and dedicated enough to use our technical knowledge.

4. The costs of risk management often are compounded by public misapprehensions of risks, usually through lack of information and through misunderstanding.

5. The results of failure in risk management are, at the minimum, what we call incidents. At the maximum, the

---

[1]The statement by Dr. Dubos is from a letter to DuPont management reprinted in a 1963 corporate brochure entitled "Chemicals and Public Health."

results are injuries, and sometimes the loss of public consent necessary for us to conduct our businesses.

Therefore, we must let the community know that management's concern extends to the community as well as to the company facility. (See Fig. 6-1.)

Community Understanding and Consent

Let us proceed now to the final aspect of our concern with risk management: How do we make the process and the results of risk assessment more meaningful to the communities in which we operate?

Plant communities need to be informed about the nature of local industrial processes and the precautions taken to minimize risks. But we do not think it is productive to dump on the general public a barrage of statistics generated by the risk assessment process. In fact, statistical projections intended to show that likelihood of disaster is very remote may only raise new alarms among the uninitiated. *Risk communication should focus on risk reduction and preventive measures rather than numbers.*

In our experience, understanding is best enhanced when the company takes the plant's safety story to the community. A number of programs can be helpful. Let us offer just three examples:

1. We must rely heavily on *presentations from managers and technical experts.* In this scenario, spokespersons go willingly to meetings of every interested group, ranging from service clubs such as Rotary and Kiwanis to schools, church auxiliary groups, and gatherings of political activists. Speakers must be prepared to answer questions candidly and fully, explaining the facts in simple language.

Risks and hazards should be realistically evaluated where skepticism or confusion persists.

Emphasis, however, is placed on what management is doing to contain and neutralize these risks. We emphasize the positive side of the story: the thorough, detailed, dedicated application of safety rules, regulations, oversight, and training. And we strive to keep it simple and in a form that listeners can pass on.

2. We can benefit from the scheduling of *open houses* at periodic intervals. Members of the community including civic officials and "thought leaders" can walk through the plant and have processes explained and safety procedures illustrated.

Families of employees are especially welcome on these occasions. Fully informed employees can be very effective in passing safety information on to their peer groups.

Confidence in the plant communities is best achieved by person-to-person contact with plant employees. They are the company's ambassadors to the public.

3. We also can benefit from the formation of *community advisory councils.* This is a relatively new concept, on the cutting edge of the plant–community interface. Members of such advisory groups can make managements better aware of community concerns and apprehensions, and can help to explain industry's story to their friends and neighbors. In some cases, such teams can be part of the decision-making process when expansions or conversions are contemplated.

The initial reception for advisory councils was so positive that the North American chemical industries adopted this strategy as part of an overall program that is called Responsible Care in the United States. We believe this is a very constructive program, but one that more properly is part of industry's response to emergencies rather than part of its communication effort.

To give another example, the Chemical Manufacturers Association prepared a brief report on the Kanawha Valley Hazard Assessment Project of the Kanawha/Putnam Local Emergency Planning Committee of West Virginia. Worst-case "scenarios" were reviewed by industry with the community on June 3 and 4, 1994. This was an outstanding community event and an excellent example of community involvement.

Let us sum up our position very briefly:

- Risk management will be increasingly challenging as technology expands to meet growing social demands. We must be equally alert in developing the skills of risk analysis and in installing the controls that will keep new technology safe.
- Risk analysis and risk assessment are highly useful, sophisticated tools, but even more important is the demonstrated attitude of management. A policy of total prevention and total dedication to safety as management's highest priority is most likely to succeed.
- We have the knowledge and the skills, and even the financial resources, to solve all the technical problems of process safety. The effort to make processes safe is ongoing because processes are continually changing.
- We still must fight a daily battle to overcome the human factors: fatigue, boredom, inattention, forgetfulness, and bringing personal problems to work along with the lunch bucket.
- Risk management means more than solving the safety problems on the plant—it includes winning the confidence, respect, and goodwill of the greater plant community. All members of the plant force should consider themselves ambassadors to their own community.

Working Relationships with Local Media

Every unit of an industrial organization should have a well-developed, well-defined relationship with local newspapers and TV and radio broadcasting outlets. Large headquarter installations usually assign these responsibilities to a public relations or public affairs division, by whatever name. Those responsibilities include communications on safety events and issues and the ability to respond promptly and accurately to media inquiries about ongoing safety provisions and actions taken in emergencies.

The CEO, however, remains the "officer in charge," and, in keeping with public relations traditions, professional spokespersons remain as anonymous as possible. At other sites, the plant manager or the branch office manager acts as the spokesperson although specific tasks may be assigned to a professional communicator or the best qualified safety professional on the scene.

Sympathetic treatment from the media at times of emergency is best assured by a policy of building up trust and friendly relations when no emergencies are creating an atmosphere of uncertainty. At quiet times, local news units are looked upon as a source of news; it is well to build such a pipeline that can be used when the occasion demands. Friendly treatment also is likely when friendly relationships are well established. For this reason, local managers should welcome opportunities to get to know media people through service clubs and other community organizations.

To put it another way: *a good working relationship with the media is a reciprocal arrangement.* It cannot be turned on and off at the whim of the company. If we in industry refuse legitimate news media requests during less than ideal news situations, or consider the requests a nuisance in the daily flow of business, we can hardly expect the media to honor our needs in emergencies.

An emergency situation also may attract out-of-town newspeople, from the state, regional, or national level, who have no interest in establishing a continuing relationship with the local operation. They will have the same appreciation as the local newspeople, however, of a good communications efforts by the local unit.

The Designated Spokesperson

Each plant or local unit should designate a primary media contact and at least one alternate who can respond to the media in the primary person's absence. Each contact should be thoroughly knowledgeable about all local operations. Each should become acquainted with area newspaper, radio and television reporters and editors, and know their deadlines, telephone numbers, and editorial philosophies.

In addition to the primary contact and the alternate, others on the plant may be thrust into the role of spokesperson because of their expertise in special fields. Such employees might include an environmental coordinator, a plant physician, or an employee relations specialist. To prepare for this eventuality, training programs ought to include on their agendas sessions on dealing with the press.

In recent years, special training programs have been developed by experienced former news editors. These programs prepare spokespersons through role-playing situations such as "facing the press." These sessions may be expensive, but they can greatly increase the confidence of a manager–spokesperson. They can even enable that person to take useful initiatives in such interviews.

Also in quiet times, the working communicator can help make a media representative's job easier by providing him or her with a packet of material about the facility, including photographs of facilities, biographies of management, and background information about products made or distributed from local facilities.

Adversarial Cautions

Generally, it is not a good idea to go "off the record" with a reporter or to try to kill a story. Many business representatives still have raw wounds from stories published with information they gave "off the record" to reporters. Professional news people, after all, make their careers by creating headlines and serving up interesting morsels to *their* customers, the reading public.

When Incidents, and/or Injuries, Do Occur

You should promptly assemble as much information as possible and phone it to the news media. Usually it will be possible to provide sufficient details on the first call to cover the entire story.

The information provided should include the time, the location, the area in which the incident occurred, the cause if known, and the extent of damage if it can be estimated. Above all, it should include the full names, ages, job titles, and home addresses of all injured persons, with some indication of the severity of the injuries. If employees are hospitalized, it also should provide the name of the hospital.

Under ordinary circumstances, the names of the injured

should be released immediately after the next of kin have been notified. Release of the names should not be unduly delayed, however, if the next of kin cannot be contacted promptly. News reports that list a number of unidentified victims can spread fear through an entire community. Management's obligation, therefore, goes beyond the families of the injured to the families of all employees.

The Bad Story

In matters of safety, as with other local issues, it is best to resist making a quick, impulsive response to an unfavorable story. Anyone in the public eye must expect some brickbats along with the bouquets.

Where inaccurate, incomplete, or distorted reporting is an issue, however, it should be brought to the attention of the media people involved. Pointing out errors or serious omissions is entirely proper. Hopefully the record will be set straight. Unless the errors are serious, however, it is advisable not to try to go to the mat with a reporter. One can seldom win an argument with people who buy printers' ink by the barrel.

Sometimes the public has a short memory. But it can have a long and painful memory if safety gets short shrift.

Figure 7-1. Plant scene showing unloader in safety suit. Multi-million-dollar investments as well as individual employees' health and well-being are in jeopardy if safety procedures are not scrupulously followed. Process down times and worker inefficiency are inevitable consequences of even minor safety incidents.

7

Safety and Productivity

Lenin is reported to have said, early in the great Communist experiment, that productivity would be the key to the success of Communism.

Any leader in the enterprise economies might well have made the same prediction. Over the long term, no society can consume any more than it can produce or exchange at market value with trading partners.

The collapse of the Soviet economy, and with it the Soviet social and military structure in the waning years of the twentieth century, proved that Lenin was right, but not in the way he imagined. There were many reasons behind the failure of Communism, but failing productivity was perhaps the most telling of those reasons.

Any nation's productivity is the sum of the productivity of all its thousands of producers. To some extent they rise or fall together. But in competitive economies, each producer must sustain a high level of productivity against others who aim at the same market. Managements fight this battle every day and must look askance at programs that, however meritorious, might undercut that productivity.

Good safety programs, properly conceived and ably administered, can make a constant, steady contribution to any organization's productivity. The relationship between safety and productivity is evident throughout this book. This chapter summarizes key aspects of the relationship.

Direct Contributions of Safety to Productivity

- Skilled employees working safely provide a constant input to the production of goods and services. (See, e.g., Fig. 7-1.) Conversely, any lost-time incident almost invariably requires the substitution of backups who are less skilled, less practiced, and less confident than the people they replace. They are also less likely to maintain a high level of safety consciousness in meshing with other employees' performance.
- Safe performance of duties minimizes the disruption of efficient operations. Conversely, any incident almost invariably causes immediate down time. And when operations are resumed, less sure hands are at the tiller, and closer supervision is required.
- Safe performance enhances the maintenance of facilities at optimum operating condition and minimizes any loss or deterioration of working materials. Conversely, incidents usually result in disrepair to facilities and degradation of materials,

at least briefly. Replacement of damaged items often adds significant production costs.

Indirect Contributions of Safety to Productivity

- Safe workers generally perform better than unsafe ones across the board. There is a demonstrated correlation between efficacy of safety training and awareness of quality control, reduction of absenteeism, and sharing of goals and objectives.
- Safe workers contribute to a generally high level of morale throughout their work groups. They tend to be more attentive, more prompt in response to orders, and better able to work with others as a team.
- Even when workers gripe about how working safely slows them down—and they frequently do—they equate management, time, money, and effort spent on safety with a management interested in them as people. Having employees believe they are being treated as people, not just as numbers, is a major step toward good morale. Good morale causes the *total* group to work toward known and desired goals, including safety, efficient production and innovation.

We firmly believe that productivity is achieved through the total exercise of good morale, with safety as the key building block.

SIKKERHEAD

SEGURIDAD

VEILIGHEID

SAFETY

SECURITÉ

SICUREZZA

ΑΣΦΑΛΕΙΑ

SICHERHEIT

SEGURANCA

Figure 8-1. The word "safety" in several different languages.

8

Safety Within Growing Global Competition

GLOBALIZATION

For nearly two centuries, the U.S. industrial establishment has been engaged in many kinds of formal and informal interchanges with its counterparts in Europe. And we in the United States are all affected by an emerging global economy that embraces the established and developing countries of Latin America, Africa, and Asia—in particular the dynamic economies identified as the Pacific Rim countries.

One of the buzz words of the 1980s, globalization has become a powerful paradigm in the 1990s. It is more than a convenient semantic specification. It is a summation of profound events taking place in the economies of developing as well as highly industrialized countries.

New technologies leap across national boundaries. And increasingly they tend to neutralize the various "economic communities"—regional combinations structured to level the playing fields, providing internal economic parities while erecting artificial barriers to trade with outsiders. The rationalization for these arrangements always has been to protect jobs, markets, profits, living standards, and even cultural heritages. There has been some merit in these protective devices, but they are eroding under new political pressures and economic necessities. Witness the NAFTA agreements and the latest changes under GATT.

Additionally, their interdependence on materials as well as on skills and expertise compels businesses—large and small—to be prepared to compete for customers in an international marketplace. Such competition obviously is more massively encountered by large corporations than by small entities. But the family farmer, the neighborhood pharmacist, the auto repair shop, and the greengrocer face international competition in myriad forms.

During the past decade, many corporations including DuPont have abandoned old mindsets that pitted home plants against "offshore" plants, home offices against subsidiary operations, and local citizens against foreign trainees. Today, a U.S. company may have a profit center headquartered in Japan, Switzerland, or Germany, depending on where the principal market or most efficient production facility is located.

These changes have not been sudden even if they often have been dramatic. They really are an escalation of a long-term grooming of the best people with the most useful know-how, wherever they have been identified. But the cumulative effect does lead to the recognition that a successful international firm must now be something subtly different—a truly global corporation.

International Aspects of Safety

The exploding technologies of the broadening industrial world tax the ability of each of the contending countries to keep pace with them. Although English today is the closest thing to a universal working language, everyday efforts are carried on in dozens of different tongues and dialects (see Fig. 8-1). Yet, however translated, safety still must embrace the same concepts and draw upon a shared heritage (Fig. 8-2).

From 1988 to 1991, author William Mottel visited Europe several times to participate in a series of safety forums established by OECD countries—that is, the countries participating in the Organisation for Economic Cooperation and Development. Ports of call included Berlin (before the wall came tumbling down), Stockholm, and London.

About 100 representatives from Japan, the United States, and the countries that make up the European Economic Community met to achieve a worldwide consensus on how to avoid catastrophic events and, when necessary, how to provide emergency responses.

The success of these meetings led to regional meetings during 1990 in Hungary and northern Spain. Having spent some years working in Europe, Bill Mottel found that those trips gave him a fresh impression of the diversity of Old World cultures. Today, there is still the constant challenge to blend these differences into a common purpose that he found during an earlier DuPont assignment in Geneva.

One humorous way to describe differences is to use special definitions of the inhabitants of Heaven and Hell as seen from varied national perspectives. The residents of heaven, by nationality, would be:

Figure 8-2. Safety record signs in Spanish and Portuguese.

French: cooks.
English: police.
Germans: mechanics.
Italians: lovers.
Swiss: organizers.

As you can see, heaven allows the "best" national traits to be properly deployed. Hell, on the other hand, is a different matter; here the inhabitants would be:

French: mechanics.
English: cooks.
Germans: police.
Italians: organizers.
Swiss: lovers.

Business managers who have carried out assignments in Latin America or in the Orient probably could suggest many other permutations on these impressions of cultural and ethnic differences.

Of course, in today's world, these old stereotypes have little meaning. Diverse organizations everywhere need, and are finding, not only good mechanics, organizers, and rule makers, but computer mavens, accountants, lawyers, and dozens of other specialists.

On the other side of the coin, whatever our individual vocational skills, we *all* need to be experts at managing safety.

The Challenge to U.S. Multinational Corporations

All the difficulties and problems companies have in creating successful safety programs in their home bases in the

United States are compounded when their offshore subsidiaries develop equivalent programs in their own countries and regions.

Each country has its own rules and regulations though administrators tend to draw on a common storehouse of information, and supranational organizations such as the European Economic Community (EEC) foster meetings, seminars, dialogues, and forums.

At times it appears that these transnational efforts add another layer of bureaucracy; at other times the establishment of common goals and specific target dates give safety efforts increased momentum. One thing is sure: safety professionals must be expert in more than safety; they must be attuned to the political crosscurrents that inevitably spawn new regulations and establish new parameters of compliance and adjudication.

Let us offer our delineation of the different mindsets that tend to come into play when we are called upon to confer with our fellow professionals in the Old World.

In Europe we see:

- Greater tolerance of bureaucracy than in the United States.
- Greater reliance on directives.
- A more conspicuous role for academic role players.
- Greater faith in elaborate protocols.
- A higher profile for union leaders.
- Greater support by governments in individual countries for private business firms engaged in global commerce.
- Greater tolerance for marketing arrangements, including cartels.

On the other hand, U.S. practitioners come to the conference table with:

- Less faith in bureaucracy than the Europeans have, and, instead, a tradition of distrust.
- Impatience to curb the avalanche of paper.
- A willingness to look to academic specialists for technical expertise, without necessarily giving them an oversight role.
- Little assurance that their own government is willing to help its own industries in international markets.
- A reluctance to join forces in any agreements or arrangements that might raise the bogeyman of anti-trust back home.

As another example, U.S. managers working abroad have difficulty in adjusting to the principle of co-determinism as it operates in certain European countries. Even firms that have had a fairly tranquil relationship with their wage roll people are nonplussed by the fact that union representatives sit at management's table when many economic and personnel decisions have to be made. These may include decisions on safety regulations.

U.S. businesses with operations in Europe must, of course, abide by the rules and protocols of the host countries. This requires a constant effort to keep abreast of not only the rules in any one country but also the differences in the rules within the European Economic Community. And it requires an awareness of political and social developments that may produce changes down the road.

For those who have not yet encountered it, the EEC is a European community composed of 12 countries, called member states. They are dedicated to the elimination of trade blocks caused by differences in national tax structures, currencies, banking codes, and so forth.

For our present purposes, this goal means the elimination of differences in safety and health standards. This effort was scheduled to be accomplished by member-ratified legislation by the end of 1992, but as of this writing, it is still ongoing.

The organizations embracing sovereign states in Latin America and in the Asia-Pacific subcontinent have not gone as far as this in developing their own protocols, but there the exchange of experience and expertise is proceeding at an accelerated pace.

Cross-cultural Parameters

In addition to differing laws, regulations, and codes, U.S. multinationals need to be sensitive to cultural differences. These go beyond ordering sushi in Japan, eating donkey meat sausages in rural France, and dining on roasted dog in the Orient. Differences also impinge on safety parameters in many countries.

Let us offer a few examples, inconsequential in themselves but illustrative of the dozens of differences that U.S. managers must accommodate.

Stainless Steel Chopsticks. An experience in Taiwan challenged the ingenuity of managers trying to find a solution within parent company guidelines. Wooden chopsticks were provided for meals served in the company lunchroom, and the long-time practice had been to recycle them after cleaning. But the temperature of the cleaning detergent solutions was found to be too low to remove every possibility of bacterial contamination, and raising the temperature to the levels needed to assure safety wreaked havoc on the wooden utensils. The solution was to provide stainless steel chopsticks embossed with the DuPont logo. These chopsticks were enough of a novelty to overcome a traditional hostility to things new—and the steel was durable through thousands of dishwashing cycles (Fig. 8-3).

Lunchtime Brew. In Northern Europe, beer and wine have long been customary lunchtime beverages. DuPont

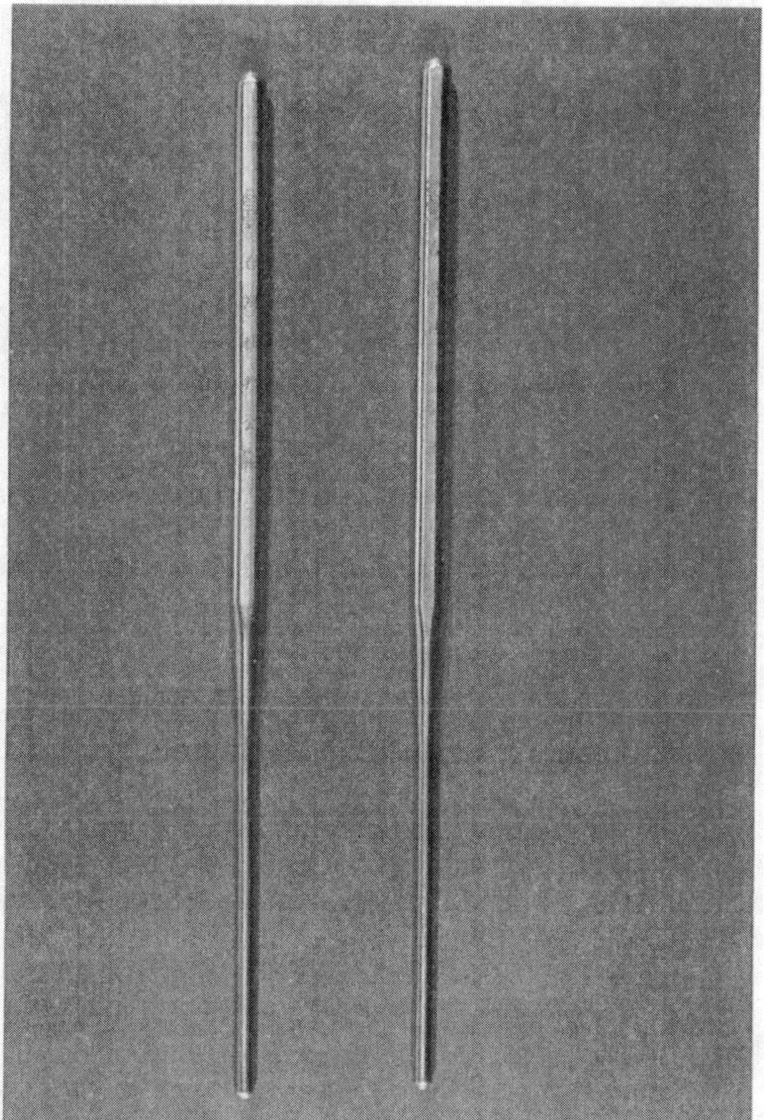

Figure 8-3. Stainless steel chopsticks replaced traditional wooden chopsticks in cafeterias on Asia-Pacific facilities. Wooden implements deteriorated rapidly under high temperature cleaning methods required to destroy all bacteria.

service managers, trained to think of alcoholic imbibing during duty hours as grounds for dismissal, were appalled. But as one German manager put it, "Milk is only drunk here by babies, and water by invalids in the hospitals." Supervision had to settle for careful oversight, while trying, with only minimal success, to wean employees toward soft drinks and fruit juices.

Window Light versus "Neons." U.S. construction methods for industrial plants have evolved to employ large windowless walls combined with well-tested banks of fluorescent lighting. Fluorescent lights provide constant light levels with minimal shadows and add a considerable safety factor to operations, not only in production and warehouse areas but in offices and laboratory areas as well. For years, however, European employees working near the few traditional windows quietly turned off the "neons," as fluorescent lights are called in many European communities, and continued to work with inadequate illumination. Management had to settle for gradual acclimatization.

Persistent Effort versus Destiny. In Middle East operations, an employee's passionate commitment to safety (determination to go the extra mile for incident prevention) sometimes comes into conflict with a resignation to fate. Although obvious safety measures are followed faithfully, incidents that occur because of a lack of imaginative foresight are sometimes shrugged off as the will of divine Providence, by whatever name. Gradual but constant insistence, however, can overcome a pervasive air of fatalism.

Insurance versus "Personal Honor." In one of DuPont's joint ventures in Japan, a company auditor discovered that a warehouse with a large volume of expensive combustibles had never been provided with a sprinkler system. The Japanese manufacturing superintendent explained that a sprinkler system was much too expensive. He gave his personal promise to "forbid that any fires would take place," saying that he would keep the hazards at bay as part of his "personal honor." Fortunately no fires occurred during the following two years, but upon his retirement, his successor ordered in a sprinkler system as an early item of business.

Spirit Houses in Thailand. No matter how thorough the safety training and procedures at an agricultural chemicals formulation and packaging plant, employees did not feel safe until a priest had blessed a newly installed "spirit house" on the site. The company accommodated itself to this tradition.

Fire Mains in the Netherlands. Dutch engineers looked askance at the insistence of U.S. engineers to run new and different tests on fire mains; as the world's leading specialists in hydraulic engineering, the Dutch engineers felt their competence was being challenged. But when the "friction loss" tests showed a severe loss on one side of the loop, a follow-up inspection convincingly revealed a gate valve with a sheared stem. An old safety adage says that one test is worth a thousand opinions.

Speed Control Stratagem at an Explosives Plant Site in Brazil. The jostling of dangerous truckloads was minimized by a 15 mph vehicular speed, caused by posting 5 mph signs. But then, drivers on roadways worldwide tend to limit their speed by driving a certain amount over the posted limit.

Global Downsizing

Midway through the 1990s, one of the greatest threats to maintaining highly effective safety programs is the worldwide wave of downsizing or restructuring of literally thousands of business firms. The effect of this dramatic change in business organizations was mentioned in Chapter 2, but it is worth noting here that all the potential problems are compounded when the streamlining has to be carried out globally. Local rules and regulations, local public relations pressures, and the reassignment of personnel, sometimes across national boundaries, necessitate including safety factors *early* in the restructuring process.

Safety Careers in the Global Organization

For safety professionals, global operations sometimes open additional opportunities for making contributions with concomitant rewards. For the person with high promotion potential, however, the prospect of safety assignments offshore may pose dilemmas. For example, will horizontal movements complicate the person's career path? Will he or she return to an organization with new faces at familiar desks, revised goals, and managers who have forgotten to think of the offshore safety manager as an outstanding resource?

Corporate planners must take whatever measures are necessary to make sure that these oversights do not happen.

Political Factors in Global Safety Operations

Effective pursuit of safety objectives in offshore countries may require special knowledge and special abilities to interface with the political environment of a host country. It is essential that employees with high level responsibility for safety have the prestige, and thus the clout, necessary to deal with governmental officials.

Summary of Global Considerations

At the beginning of this chapter, we had a little fun with cross-cultural differences. Let us finish with another list that highlights the difficulties of rule writing and rule enforcement:

- In Britain, everything is permitted that is not forbidden.
- In Germany, everything is forbidden unless specifically permitted.
- In Switzerland, everything that is not forbidden is compulsory.
- In France, everything is permitted whether forbidden or not.
- In Italy, everything is permitted in love and forbidden in war.

In the United States, we live and work in a different society, a melting pot with a very cosmopolitan tradition. So we have to deal with an even more difficult cultural situation:

- In the United States, everything is permitted somewhere, forbidden somewhere else, and argued about almost everywhere by someone or other.

About safety, however, there can be no argument anywhere in the world. Safety and profitability go arm in arm. Safety and profitability and cost-effectiveness are everywhere important, and the only restraints we have are those of unequal skill and experience.

The human relations aspects are especially important in global operations. We learned in our international service with DuPont that the real and lasting success of an operation depends, as it does in the United States, on building mutual confidence, respect, and trust among all employees. Wherever those of us who are associated with multinational companies work, the success of our programs ultimately depends on this.

Figure 9-1. The Cray XMP super computer was one of the most modern and powerful computers in the world when installed at the SRP site. It served both the Savannah River Plant and the Savannah River Laboratory. Its ability to perform 300 million operations per second made feasible the extremely complex calculations needed for nuclear reactor accident safety analyses.

9

Nuclear Energy—Special Problems, Special Solutions

The special safety problems and issues in the development and use of nuclear energy are not widely understood. In part this lack of knowledge is due to the new and radical technologies involved. It is also partly due to national defense requirements that shrouded most nuclear activities in secrecy for many decades.

It is possible that the U.S. government and managements, by playing the nuclear cards close to the vest, delayed the development of nuclear expertise among potential enemies. Gaps in public knowledge, however, have contributed to much misinformation and much misunderstanding. The inherent dangers in radioactivity, for example, have spawned frenetic behavior among some activist groups. Many problems still to be resolved, such as the safe disposal of nuclear wastes, generate more alarm than necessary and cloud the nation's nuclear policies.

To shed some needed light on this subject, in the following pages we discuss in some detail the known aspects of DuPont's nuclear safety effort, stressing a number of truly remarkable accomplishments.[1]

DuPont's Nuclear Involvement in World War II

DuPont became involved in the production of nuclear materials for national defense with the beginning of the Hanford project, under the Manhattan District of the U.S. Corps of Engineers, in World War II.

By the time of that war, DuPont had become a highly diversified manufacturer of chemicals. Only one of its several departments produced explosives. For the World War II nuclear program, DuPont had little directly applicable expertise to offer—only great professional and managerial competence. The key to its participation was the willingness of the company to do its very best when drafted into its country's service. Its best people were selected, and they were committed to the effort. Profit was not a motive.

However, DuPont had not gone into the Hanford project completely blind. Although it had no nuclear experience, several of its top scientists and engineers, including Crawford H. Greenewalt, had been members of a reviewing committee to advise on nuclear development, particularly on the "Metallurgical Project," the cryptic designation applied to Arthur H. Compton's work at the University of Chicago. Greenewalt was present on December 2, 1942 when, under the general direction of Enrico Fermi,

---

[1]This chapter draws extensively on the *History of Du Pont at the Savannah River Plant*, written by Dr. William P. Bebbington, who spent most of his career working at that facility. Copies are available from the Aiken, South Carolina Chamber of Commerce.

a self-sustaining chain reaction was first achieved in a uranium–graphite "pile" under the Stagg Field stands.

A feasibility report addressing the engineering aspects of a production plant had been prepared. When DuPont took over a major part of this project at the end of 1942, however, no site had been chosen, and important questions regarding reactor cooling had not been answered. There were no nuclear engineers and only a handful of nuclear physicists, chemists, and chemical engineers; most of them were already fully occupied with experimental work.

Despite these sparse resources, by early summer of 1945 three Hanford "piles" were operating at designed power, and finished plutonium was being delivered to the next phase of the Manhattan Project. The first plutonium test device was exploded in New Mexico on July 16, 1945. The rest is history.

It is not the purpose of this book on industrial safety to rehearse the full development of nuclear energy. But it needs to be noted that the absence of extensive experience in the production of nuclear materials also carried with it the absence of any sure knowledge of the complex safety problems to be met in this pioneering journey.

After the war, DuPont ceased its operation of the plant at Hanford as it had previously agreed to do. The company had no aspiration to be part of a commercial activity to produce and market power derived from nuclear energy. However, several DuPont specialists worked closely with government officials in determining what developments in nuclear science and engineering were critical to national defense.

The Savannah River Project

In July 1950, U.S. President Harry Truman sent a letter to Crawford H. Greenewalt, who had become the president of the DuPont Company. The letter (Fig. 9-2) was reminis-

THE WHITE HOUSE
WASHINGTON

July 25, 1950

Dear Mr. Greenewalt:

The Atomic Energy Commission has informed me that it has requested the DuPont Company to undertake the design, construction and operation of certain new facilities for the atomic energy program.

The Commission advises me that the Company has within its organization technical, scientific, engineering, construction and operating staffs capable of handling a task of this magnitude. The great resources of your Company in these fields, together with the experience which it has acquired through the successful handling of the design, construction and operation of the Hanford Project during the War make it uniquely qualified to undertake this most essential task.

I want you to know that I consider this project as one of highest urgency and vitally important to our national security and defense.

Very sincerely yours,

Mr. Crawford H. Greenewalt
President, DuPont Company
10 Market Street
Wilmington, Delaware

RECEIVED
JUL 26 1950
C. H. GREENEWALT

Figure 9-2. Truman letter to C. H. Greenewalt.

cent of the letter sent to E. I. du Pont by Thomas Jefferson almost 150 years earlier. It conveyed the request of the Atomic Energy Commission that DuPont should "undertake the design, construction and operation of certain new facilities for the atomic energy program."

Intense study was given this request by company committees bringing together DuPont's top management and specialists who had gained extensive experience at Hanford. A letter from Greenewalt to President Truman (Fig. 9-3), dispatched in October, confirmed DuPont's willing-

DU PONT
ESTABLISHED 1802
E. I. DU PONT DE NEMOURS & COMPANY
INCORPORATED
WILMINGTON 98, DELAWARE

EXECUTIVE OFFICES

October 17, 1950

THE WHITE HOUSE
OCT 18 8 ?? AM '50
REC'D

The President
The White House
Washington, D. C.

My dear Mr. President:

I have long delayed acknowledging your letter of July 25 in which you refer to the new project the du Pont Company has been asked to undertake for the Atomic Energy Commission. I have delayed this acknowledgment because I wished to be able to tell you when I wrote the outcome of our discussions with the Commission.

We have today executed a Letter Contract with the Commission to cover the design, construction and operation of these new facilities. As was the case with the Hanford project in World War II, we have elected to undertake this new assignment on the basis of cost plus a fixed fee of one dollar. I hope that we will be able to justify the confidence you were kind enough to express in your letter to me. In any event I can assure you that the du Pont Company will, as always, put forth its best efforts.

x/190

Very sincerely yours,

C H Greenewalt

C. H. Greenewalt
President

CHG:NBS

Figure 9-3. C. H. Greenewalt letter to President Truman.

ness to go ahead on the program vital to the country's security.

By February 1951 construction was under way at a site alongside the Savannah River, which flows between the states of South Carolina and Georgia. At the time, that undertaking was the largest single construction project in the history of the free world. At the same time, the assembly of an operating force had begun.

Nominal production rates for the first production units were achieved in April 1953. Full production was achieved in 1955.

Safety Programs at Savannah River

DuPont's tradition of protecting its employees from harm at their workplaces was emphasized at the Savannah River Plant, and the Savannah River Laboratory which was an integral part of it. Special attention was given to the unusual hazards of working in proximity to sources of radiation and with radioactive materials. DuPont standards for radiation dose at Savannah River were set 40 percent below government limits. In over 35 years of operation only two employees exceeded the federal limit. In the final 15 years of DuPont's involvement at Savannah River, no employee exceeded the DuPont limit.

Of constant concern to management and to reactor engineers and scientists were the risks associated with reactor operation and ways to contain the risks and mitigate the effects of a safety incident, should one occur. Management recognized that as power levels were raised, so too were the consequences of potential incidents.

Efforts in these directions were continuous and were increasingly effective as understanding of the physics of heavy-water reactors increased and deepened.

Safety was greatly enhanced by the tremendous advancements made in electronics and computers. These new technologies were quickly adopted and adapted as they became available. (See, e.g., Fig. 9-1.)

Equally important, DuPont took a leading role in the addition of devices such as cooling sprays, filters, and auxiliary sources of cooling water, as well as the constant scrutiny and revisions of operating limits and procedures.

In 1956, a supplementary safety system was installed so that a reactor could be shut down by the operators if the control and safety rods failed to drive or drop into the reactor. Originally a safety procedure had been provided in the form of tanks into which a sufficient portion of the moderator[2] could be rapidly dumped if the rods failed to go in. The new system, which was extensively tested, provided means for very rapidly injecting a neutron "poison," gadolinium nitrate, under nitrogen gas pressure.

In 35 years of DuPont operation, Savannah River reactors never had an incident that severely damaged equipment, seriously injured an employee, or released significant radioactivity to the environment.

The radiation exposure at Savannah River was far below that of the nuclear power industry. There is no evidence that occupational exposure to radiation caused cancer among plant employees. The general industrial safety record was similarly good, with SRP's record being far better than the records of other Department of Energy sites and the nuclear industry generally. (See Fig. 9-4.)

---

[2]A *moderator* is defined as a substance, such as graphite or heavy water, used to slow down neutrons from the high energies at which they are released in fission to lower energies more efficient in causing fission.

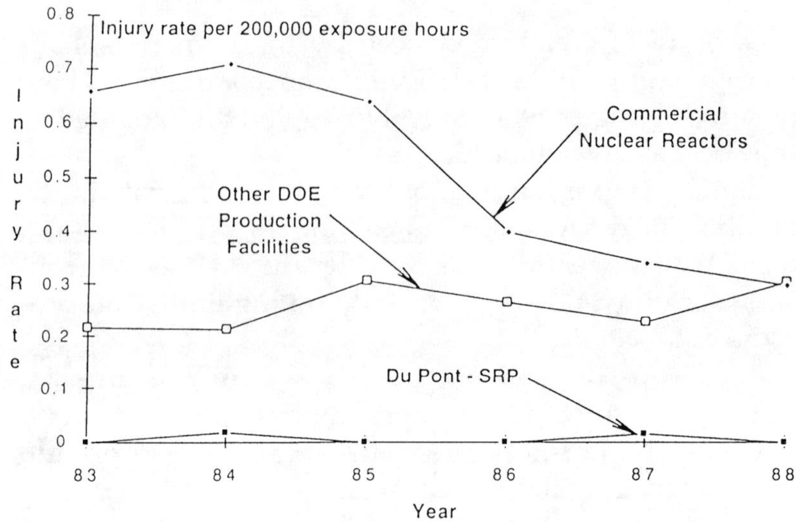

Figure 9-4. U. S. nuclear industry injury comparison.

Radiation Protection at Savannah River

The unusual and most critical hazard to be guarded against was radioactivity. Employees had to be protected from many kinds of radiation: some types of radiation can penetrate thick and dense structural materials, whereas other types are most hazardous on contact or if ingested into the body. Numerous health and environmental protection activities were undertaken at SRP (see Fig. 9-5 for some examples).

Technical protection against radiation hazards was primarily the responsibility of the health physicists within the plant's Health Physics Division, later designated the Health Physics Department. The first and most routine duty of the health physics technicians was to monitor workplaces continually. Refined monitoring ensured that barriers against radiation and escape of radioactive materials remained intact; the intensity of radiation behind the barriers could change with changes in process conditions. No employee at any level was permitted to rely on the assumption that all was well.

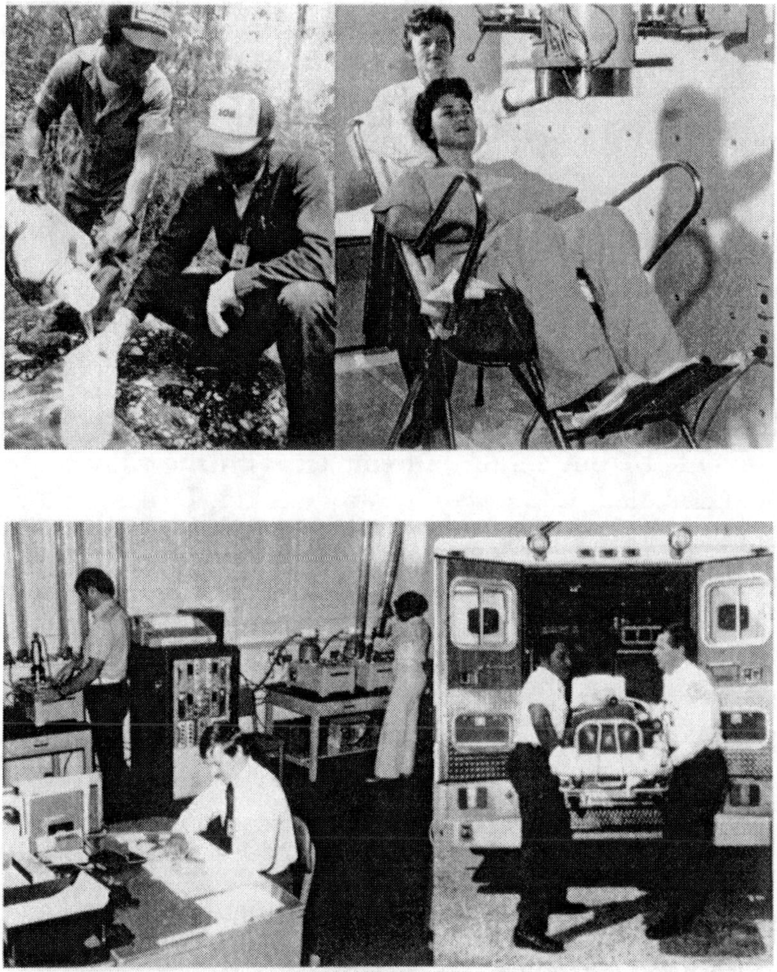

Figure 9-5. Mix of activities of the health care sector at Savannah River.

In the early years of operation some techniques were rather simple but nonetheless effective. For example, routine monitoring for radioactive contamination was done by wiping surfaces with pieces of absorbent paper, or "swipes," that were then taken to a nearby laboratory and "counted" by a suitable radiation detector. Each decomposition of a radioactive particle on the swipe resulted in an emission that was detected and registered. The relative contamination was expressed in counts per minute. Later, the manual swipes were supplemented by constant air monitors that drew air from the workspace through filters that could then be "counted."

Defense in depth, a basic DuPont safety principle, was also basic in radiation control. It began with the design and the construction of the process building and its facilities. Nuclides that emit penetrating radiation—gamma rays and neutrons—must always be kept behind thick shielding made of suitable materials. Nuclides that emit less penetrating radiation—alpha and beta particles—must be prevented from coming in contact with the skin of, or being inhaled or ingested by, any worker. To ensure that the barriers were intact, appropriate radiation monitors were provided—both continuous monitors that recorded radiation levels and sounded alarms and portable monitors, called dosimeters, that could be worn by workers, or used by them as they worked or left areas where the failure of a primary barrier might occur.

If the risk of such failure was appreciable, the workers used protective garments of cloth, plastic, or rubber and, in some cases, respirators or even complete plastic suits to which clean air was supplied through hoses. These latter requirements were most common when direct maintenance work was done on equipment known to be contaminated. Also, in such situations trained health protection technicians supervised and made their own measurements and set limits of exposure time.

People working in areas where contamination was possible removed their outside personal clothing in special change rooms and replaced it with specified protective clothing before entering the work area. The protective clothing was donned in an adjacent room before the workers proceeded to their work stations.

The basic protective garment was a zippered cloth coverall. Cloth "booties" and one or more pairs of rubber or cloth gloves were put on and sealed with masking tape to the legs and arms of the coverall. A cloth cap prevented contamination of hair. In some cases two layers of such protective clothing were required. On leaving the workplace, workers removed the protective clothing before standing on paper "stepoff" pads. In some situations, health protection technicians monitored workers with handheld instruments before they were allowed to pass back into the clean portion of the change room.

Finally, before leaving the change room, workers monitored their hands and shoes to ensure that they were not carrying any radioactivity into clean areas of the building. Their freedom from contamination was confirmed by additional monitors as they left the building. At first these final "hand and foot monitors" required that each person stop, stand on a sensing element, and put his or her hands in sensing chambers. This caused delays, particularly at the time of shift changes; so walk-over monitors were developed and installed (Figure 9-6 shows some of the monitoring devices.)

The overall radiation exposure—as opposed to removable contamination—of employees was determined and recorded by means of "dosimeters," devices that measured accumulated radiation such as that which escaped through shielding barriers. For work in zones known to have higher than normal radioactivity, employees were given, on entry, self-reading dosimeters that they could observe from time to time to ensure that they had not exceeded the permitted dose. In addition, all workers who might be exposed to radiation routinely wore dosimeter badges that were collected at regular intervals by the Health Physics Division. The exposure registered on the badges was measured and entered into employee records of cumulative exposure. In early years these personal

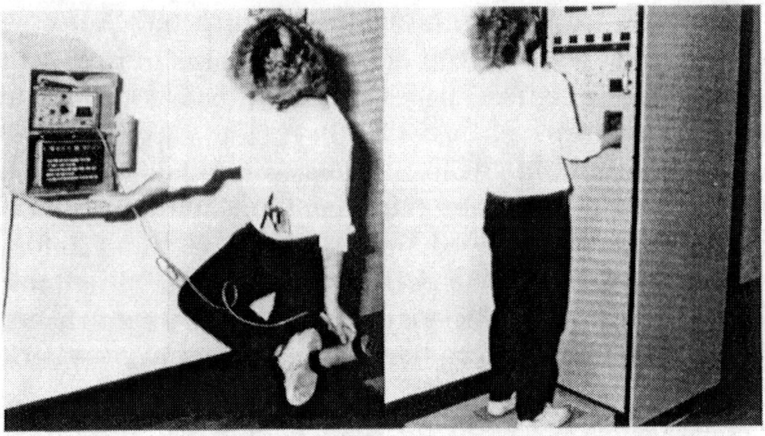

Figure 9-6. Photos of hand and foot monitors.
Primary Hand and Foot Monitor (left). Before leaving the immediate work area where radioactivity may be present, each employee monitors hands, feet and, if necessary, clothing and hair at stations such as this. Radiation dosimeters can be seen hanging from the chain to which the employee's ID badge is attached.
Secondary Monitor (right). These monitors are located at all exits from buildings in which radioactive materials are handled and everyone leaving the building must monitor himelf, regardless of whether he has been in a process area of the building. An alarm sounds if radioactivity is detected.

dosimeters were "film badges," each of which held a piece of X-ray film. The extent of darkening of the film was a measure of radiation exposure.

In 1973, Savannah River was one of the early sites to change to a more effective and quickly read device, the thermoluminescent dosimeter (TLD). In this device the sensing element is a small chip of lithium fluoride. On heating after exposure to radiation, the chip emits a burst of light, the intensity of which is a measure of exposure. SRP engineers developed an automatic reader for these dosimeters that removes the chip, heats it, records the indicated dose, and then replaces the chip in the badge. The TLD offered the added advantage that heating to a higher temperature produced a second burst of light, also proportional to the radiation dose. This provided a backup if the normal reading was missed or was far out of the expected range. Savannah River physicists also developed a variant of the TLD that recorded exposure to neutrons (the TLND).

Another innovation by DuPont at Savannah River was the inclusion of special radiation dosimeters in the security identification badges of all employees. These consisted of small pieces of indium foil that would be activated by the intense burst of radiation that accompanies criticality. These badges were always worn by everyone within the plant, employees and visitors. The dosimeter would record any unexpected and possibly severe exposure even if the wearer had not been issued the regular dosimeter badge.

Figure 9-7 shows remote controls used for handling radioactive materials, another type of radiation protection used for some of the workers.

Radiation Exposure Policy and Experience

There was never an unplanned criticality or chain reaction at Savannah River during DuPont's tenure, and to the best of our knowledge there has been none since, although the present operation has largely been on a standby basis. Also, radiation doses to the surrounding population have been minute (see Fig. 9-8).

DuPont was always conservative in its limits for radiation exposure. During most of its period of operation of SRP, the federal limit for whole-body occupational exposure was 5 rem[3] per calendar year. The DuPont limit was

[3]The effect of radiation is expressed in *rem*—Roentgen equivalent man-units. One rem is defined as a dose of any type of radiation causing a biological effect that is equivalent to that of one roentgen of X-ray of gamma-ray radiation.

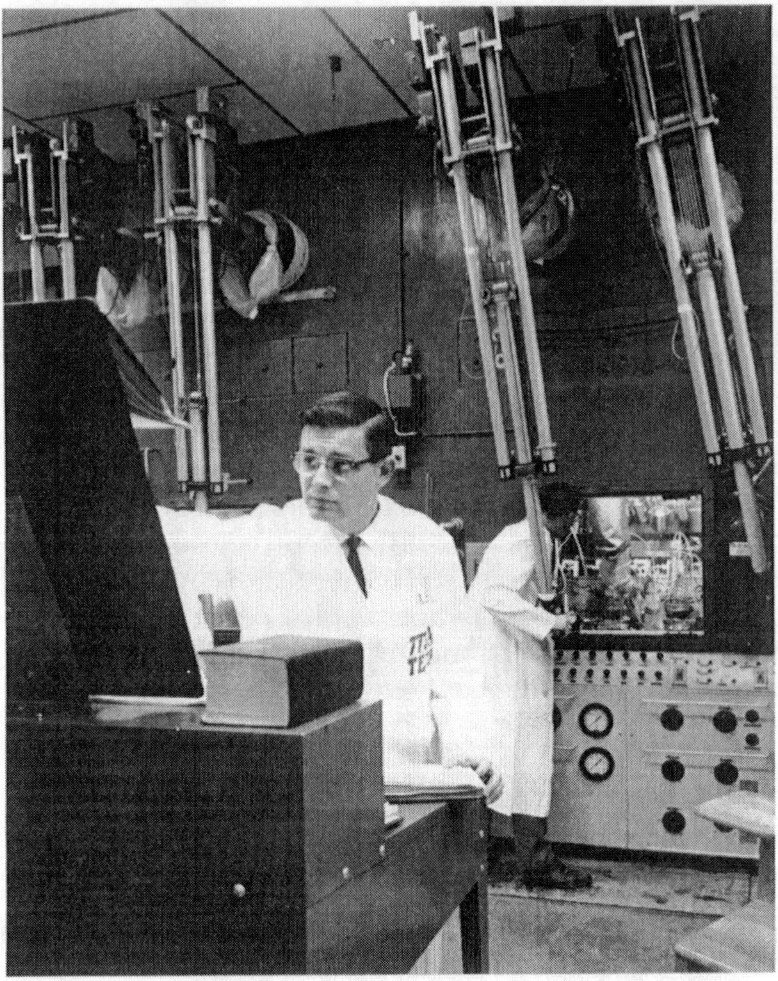

Figure 9-7. Remote controls for handling radioactive materials.

always 3 rem per year. Over 30 years ago, a worker in one of the tritium facilities at SRP experienced inhalation exposure of slightly more than 10 rem. An employee of a Construction Division subcontractor doing X-ray examination of welds also received about the same exposure from an X-ray machine. In one other instance a film-badge

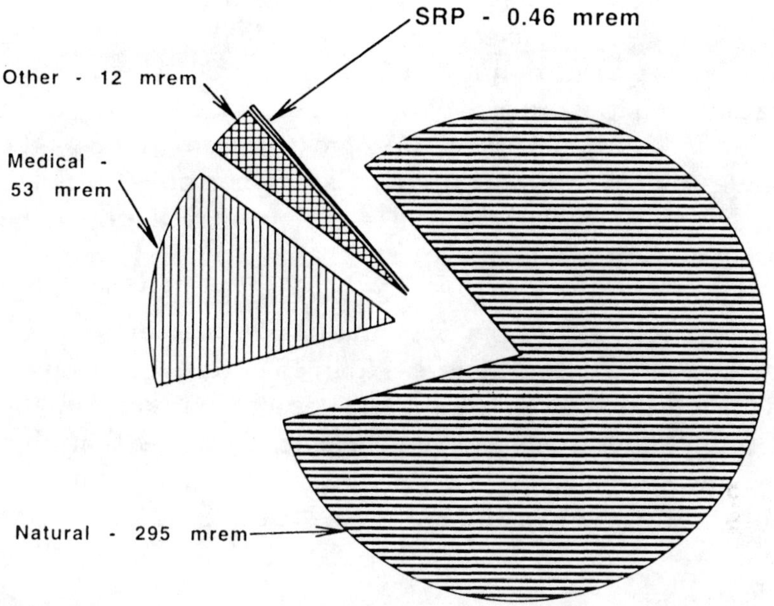

Figure 9-8. Chart showing SRP contribution to annual radiation dose of the population in the Central Savannah River Area (area near the plant), in 1988.

reading indicated that a DuPont employee had been exposed to more than 5 rem, but eventual careful review proved that his exposure had been less than the limit.

After 1975 there was no exposure greater than 3 rem during DuPont's tenure at SRP, and the trend was almost steadily downward, with a maximum exposure to one construction employee of 2.04 rem in 1988. The maximum exposure to operations employees was 1.59 rem. Single exposures of these magnitudes produce no clinically observable effects, and the calculated excess cancer mortalities resulting from such exposures are about 1 percent or less of those from natural sources.

In other Department of Energy (DOE) installations, unfortunately, there have been 3100 exposures in excess of 5 rem in a year; at nuclear power plants in the United States there have been more than 700 since 1980. The average radiation exposure of SRP workers declined steadily from slightly less than 0.5 rem in 1972 to 0.066 rem in 1988—the lowest since plant start-up. This is less than one-fourth the dose an SRP worker receives from natural radiation in the surrounding residential area.

From another statistical perspective, in 1988 the average collective dose per Savannah River reactor was 41.7 person-rem. This dose is less than 15 percent of the average collective dose of nuclear power workers in the United States and has been consistently lower than that for any known operating production or commercial reactors in the free world. Figure 9-9 provides a global overview.

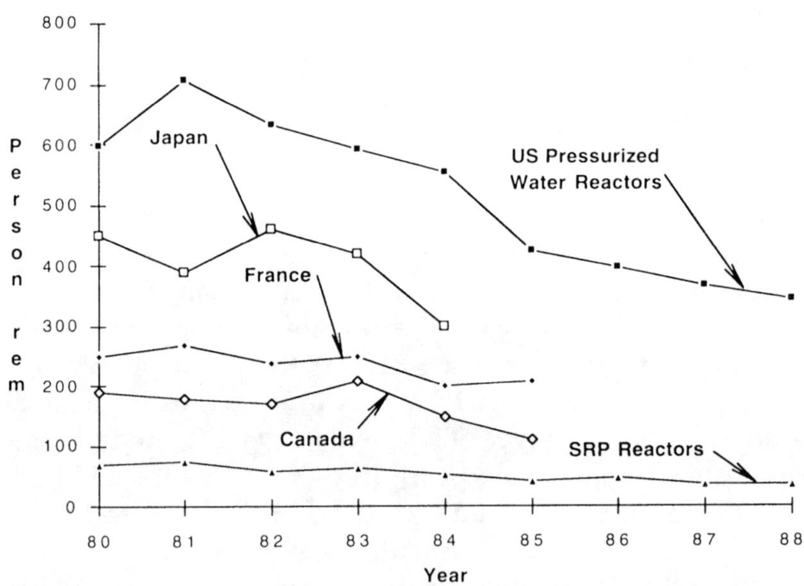

Figure 9-9. Average collective dose per operating nuclear reactor.

Assimilations of Radionuclides

Of particular concern at SRP was the possibility that radioactive nuclides could enter an employee's body; for example, by breathing contaminated air. To monitor this potentiality a program of routine analysis of urine samples was in effect. If ingestion of plutonium was suspected, fecal samples also were analyzed.

Another potential route for plutonium to enter the body was through a wound made by a plutonium-contaminated object, rather typically the sharp needle of an analytical sampler. In these cases a wound monitor was used to detect the presence and the exact location of the plutonium. Guided by the monitor, a physician in the Medical Department could selectively excise bits of tissue until plutonium was no longer detectable.

The maximum amount of plutonium that federal regulations permit a worker to accumulate in the body—the permissible body burden—is 0.6 microgram, about the size of a typical airborne dust particle. This corresponds to an annual dose to the lungs of 15 rem. The permissible bone dose is 30 rem per year. Plutonium assimilations are included with other radiation in computing the whole-body doses received by employees. In the atomic bomb development at Los Alamos during World War II, a number of workers far exceeded this limit; however, there have been no reports that these workers have suffered any ill effects in the ensuing half century as a result of those exposures.

In no case did an employee receive more than the permissible whole-body dose of 5 rem or the bone dose of 30 rem in a year. All Savannah River employees who have measurable amounts of plutonium in their bodies are listed in a national plutonium registry so that their health can be continually evaluated, even after they have left the plant and the area.

Tritium is the other nuclide of concern with regard to assimilation. As mentioned, one employee received a 10 rem whole-body dose in a calendar year as the result of tritium inhalation, without ill effects.

Criticality or Chain Reaction

SRP has never had a nuclear criticality incident. During the years that SRP was operated by DuPont, about 15 criticalities occurred in government plants and one privately owned nuclear plant. These incidents caused four deaths and over 30 exposures of nearly 500 rem.

The record of SRP with regard to unplanned criticality deserves special emphasis. Criticality or a chain reaction can occur whenever enough fissionable material such as plutonium-239 or uranium-235 is brought together in the right configuration to cause the reaction. It can be prevented by dispersal of the fissionable material, either as a dilute solution, or geometrically, by spacing or shape. Chain reactions are more difficult to achieve in thin sections than in blocks or spheres. Neutron absorbers, or "poisons," can be effective in preventing criticality.

Large amounts of uranium-235 were processed into fuel elements in one dedicated area at SRP. Fuel assemblies of this material were stored in the reactor buildings, as were irradiated fuel elements that could contain either uranium-235 or plutonium-239. In the chemical separations processes in two production areas, both of these fissionable materials were present in large quantities in solutions and, in the case of plutonium, as solid metal.

The perfect nuclear criticality safety record at SRP was the result of a very comprehensive preventive program established by DuPont. And this comprehensive program drew upon many decades of the strongest possible safety tradition, developed over those same decades in the company's commercial operations.

Management wanted no unplanned criticality, not only because of the hazard to employees—damage to equipment from a criticality is typically slight—but also because such an event would indicate failure of the procedural system. A high level Criticality Review Committee was established with the general superintendent of the plant's production departments as its chairman and the general superintendent of the plant's technical assistance department as its secretary. The director of the Physics Section of the Savannah River Laboratory and a delegate from the Wilmington Management Staff of the company's Atomic Energy Division were also members of the committee.

Effects of Radiation on Employees

Under DuPont's stewardship, we know of no Savannah River employee who ever was injured or had his or her health impaired by radiation or radioactive contamination. The incidence of cancer among plant employees from 1956 through 1974 was less than 75 percent of that which would have been expected from a properly comparable sample of the U.S. population. The overall mortality rate of Savannah River employees was 20 to 30 percent below the national average.

Comparable data for the past 20 years are not available. It must be realized that the epidemiological studies necessary for such comparisons are extremely complex. And statistically comparable numbers of cases are very difficult to find. However, we know of no specific data involving Savannah River employees that would indicate that any actuarially significant increases in injury or mortality rates have occurred.

General Industrial Safety at Savannah River

The Savannah River operations, in addition to being subject to the unique hazards of processing large quantities of a great array of radioactive materials, had all of the general hazards of a large industrial plant. The most severe non-radioactive process hazard was in the operation of facilities for the production of hydrogen sulfide gas. Hundreds of tons of this very toxic gas were manufactured and used under high pressure without a serious or a fatal injury.

Employees had to be protected from all the more prosaic injuries that could arise from falls, sharp objects, rotating machinery, high temperatures, chemically toxic substances, and—more than on most industrial sites—railroad trains and automotive equipment.

The principles established and used at DuPont commercial plants were applied at Savannah River. The principal elements of process safety management as described in Chapter 4 were well in place, particularly the management of change. Operational safety was a prime responsibility of plant management. Appropriate shares of that responsibility were delegated down through the several levels of supervision. Safety engineers were provided in each of the operating areas, but these people acted as technical advisors and monitors; line managers were the responsible heads of the safety effort.

Employees were provided, at no cost to them, with safety glasses, steel-toed safety shoes, appropriate gloves, and hard hats. These protective devices also were available for purchase for use off-plant; prices were low, and purchase was positively encouraged. As in DuPont commercial operations, incentives in the form of safety awards of considerable value were given to all employees when the plant attained prescribed levels of occupational exposure without an injury that caused an employee to lose time from the job or to suffer a permanent disabling injury (see Fig. 9-10).

It is ironic that, despite all these efforts, one of the difficult tasks of supervision was to persuade each and every employee to use his or her safety equipment and to obey safety rules without fail—a problem common to all industry.

The Savannah River Safety Record

In 1970, the Savannah River Plant received the Atomic Energy Commission's "Best Ever" Award in industrial safety for having accumulated the largest number of injury-free man-hours every achieved by an AEC contractor over five consecutive years without a lost-time injury.

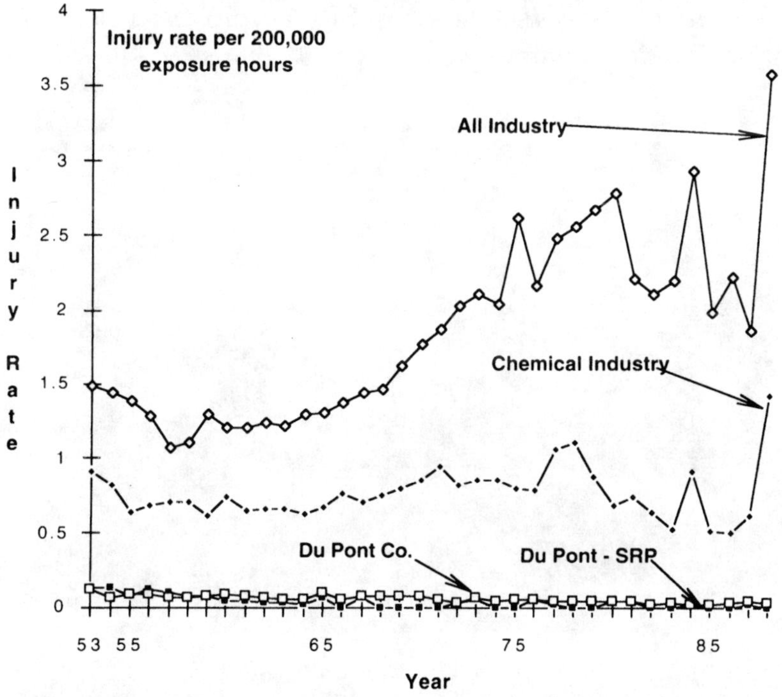

Figure 9-10. U.S. industry lost-workday injury comparison, showing Savannah River's excellent record.

In March 1989, the plant won its 35th Board of Directors award. It had operated 596 consecutive workdays—nearly 26 million work hours—without a time-losing injury.

In 1987, an injury terminated a run of 43 million exposure hours without injury. If the plant performance during that period had simply equaled the average U.S. industrial safety experience, over 600 of its employees would have sustained lost-time injuries. During the last eight years of DuPont's operation of the plant, there were six years during which there was no lost-time injury.

There were eight fatal incidents to operating employees during the nearly 38 years that DuPont operated the plant. Ironically, two of those fatalities were due to falls during an unprecedented snowfall in 1973. Over the same period there were 77 lost-time injuries.

Figure 9-11 shows a group of SRP employees celebrating their safety achievements.

Figure 9-11. SRP group celebrating their safety record. DuPont managements welcome such opportunities to celebrate special safety achievements, which help to build morale and provide employees continuing motivation to excel.

10

Safety and the Environment

Although the focus of this book is on industrial safety, it should be noted that safety and environmental protection are close cousins and in many organizations are administratively cross-linked.[1]

Many corporations have found it prudent to incorporate safety, occupational health, and environment affairs in one comprehensive effort. Excellence in any one of these areas can be made to carry over into the others.

As with safety, supervision at all levels must understand that the environment can never be compromised in the pursuit of efficiency, product improvement, market

---

[1]The material in this chapter is drawn from a speech by Dr. Bruce W. Karrh, DuPont vice-president for Safety, Health and Environmental Affairs, delivered at the Thirteenth Anniversary Joint Business Conference of the USA-ROC Economic Councils in Hawaii in November 1989.

growth, and corporate earnings, fundamental as these objectives certainly are.

The environmental impact must be carefully considered in product development and process design.

All hazardous materials, including waste, must be handled safely and properly. (See Figs. 10-1 and 10-2 for examples of waste handling at two DuPont facilities.)

The Global Factor

As noted, industries on all continents have found it necessary to compete in global marketplaces where a premium is placed on improved products and services. The necessary role of advancing technologies complicates the re-

Figure 10-1. Settling tank at a DuPont plant helps to remove contaminants from waste streams so that they can be treated or disposed of separately.

Figure 10-2. Waste treatment facilities of large plants can involve expenditures in the tens of millions of dollars. This wide-angle photo shows a settling pond, part of the wastewater treatment system at DuPont's Chambers Works in New Jersey.

quirements of safety and environment protection but does not lessen the urgency of these considerations.

Some developing nations may not be ready yet to put greater stress on clean air and clean water than on full employment, especially in the wake of recessions across many national economies, but environmental issues continue to draw vigorous public support.

Waste Management

The current focus of much environmental concern is hazardous waste disposal, to the point that waste management has become a shorthand measure for overall environmental stewardship. Some people, in fact, define all pollution as waste.

DuPont has developed a written waste handling policy that focuses on waste minimization and safe waste disposal.

Hazardous waste minimization is good business. By the beginning of the 1990s, DuPont's cost for handling and disposing of wastes had passed $100 million a year, not counting the loss of yield of useful products. And that figure was escalating at a double-digit percentage rate.

Technically, hazardous waste minimization can be accomplished by improving the yield of a process, by recycling by-products, by substituting other materials for hazardous materials, or by modifying operating procedures.

No matter how hard a company works at waste reduction, however, some waste inevitably is created and requires safe disposal. The strong preference is to destroy the material and to do so on the same site. Beyond this, society is best served by maintaining a variety of disposal options.

Incineration is one of the preferred methods. Industry is constantly upgrading the size and the efficiency of its incinerators.

Treatment of wastes by neutralizing agents renders effluents relatively harmless and sometimes creates a marketable by-product.

Landfills have had a murky history but, when properly sited and carefully prepared, can dispose of waste safely.

Transfer of wastes to other sites where the wastes pose a decreased threat or long-term environmental burden is a last-choice solution. At the present time, the traffic in waste-disposal vehicles on public highways has magnified the need for close monitoring of equipment and for vigilance in making sure that only approved sites receive the burden.

Former DuPont Chairman E. G. Jefferson once said that

if you tour an operation that has demonstrated good, sustained safety performance, you will usually find the added benefits of good housekeeping, good product quality and high morale. The linkage between safety, health and environmental matters is clear. The healthy employee is going to be more alert and more productive than one who has to cope with preventable physical problems or excessive stress. And protective environmental quality is in many ways an extension to our community neighbors of the same concerns we have for the health and safety of our employees.

11

Conclusions—A Management Guide

Two centuries of experience underscore the DuPont principle that *safety is not only a moral and ethical imperative but good business as well.* We are convinced that thousands of other safety professionals in other business firms are equally dedicated to safety.

As a parting exhortation we would like to offer Ten Commandments of Safety, with some appropriate comments on the way:

1. *Require safety as a total commitment,* beginning at the very top and carried through to the lowest-level jobs in the organization. Communications must be clear, vertically and horizontally, with respect given to feedback from those who deal with safety every hour of their working day.

2. *Build an environment of prevention through an immediate response to incidents,* not by reacting to injuries after they have happened.

3. *Recognize people as the most important element of any safety and occupational health program.* Schedule periodic training in safety procedures and practices for all personnel. Define accountability and responsibility for every member of the organization. Establish clear performance standards; reward good performance through a formal structure of group awards.

4. *Make safety the first item on the agenda of every scheduled meeting at all levels of the organization.*

5. *Root productivity in safety.* These two objectives should never be looked at, consciously or unconsciously, as competitors for time, resources, or personal loyalty.

6. *Focus collection of data and other measurement devices on identification of problem areas.* Data accumulation must not become important as an end in itself.

7. *Achieve the maximum.* Never be content with just meeting legal requirements if you can find a way to make a situation better.

8. *Share safety lore and information with all; borrow innovative ideas wherever they can be found.* Network with other departments, firms, trade associations, and governmental resources.

9. *Use outreach to improve safety off-plant*—in employees' homes, in the office or plant community, with vendors and customers.

10. *Make career paths promising for safety people*—as promising as for any professional in the organization and for professionals in trade groups and in universities.

Appendixes

A

Chronology of DuPont Safety Events

The following lists demonstrate that safety requirements can be built, "brick by brick"—resulting in a strong safety matrix over a period of time. This particular document, from the early 1980s, combined historical material with actions taken at one small, recently developed plant site at Glasgow, Delaware. The combined list in effect encourages the employees at a relatively new site to tap into a long and honored tradition.

Today, of course, safety professionals stand on the shoulders of their predecessors and, with rapid communications techniques in hand, can build a formidable safety program in a very short time.

Rules and Practices

The establishment of safety rules and practices has been a TRADITION at Du Pont since its first decade. Listed below is a sample of safety rules that have been established with prevention of accidents in mind. This TRADITION of prevention is carried on today as new rules and practices are established at Glasgow based on serious incident investigations and routine safety inspections.

A direct benefit of this practice for Glasgow employees is a safe working place which has prevented 400 lost workday injuries (as compared to industry averages) thru December 1982.

1806—Management gives personal instruction on rolling powder kegs to avoid dropping them.

1811—Copy of earliest written safety rules.

1835—Workers required to ride outside Conestoga wagons for rapid escape when transporting explosives.

1900—Standard operating procedures introduced.

—Recognized significance of individual attitude on safety performance and the need for a proper attitude in order to avoid accidents.

1904—"Strike Anywhere" matches outlawed all over the company; practice continues today.

1912—Company develops design standards for construction.

1919—Safety rules printed and posted.

1939—Ladder safety rules established.

1942—Lifting regulations established.

1945—Lock & Tag Procedure initiated.

1966—Process Hazards Review Program initiated following Louisville Plant explosion.

1980—Glasgow Site specifications for shelving erection established.

1982—*First* Site-wide Glasgow Safety Manual.

Protective Equipment and Facilities

TRADITIONALLY, the DuPont safety program has been expanded to include facilities designed for safety and protective equipment. Listed below are several examples that have been incorporated into the safety program to provide a physical barrier between a hazard and the employee. These examples demonstrate the effort to make the Company, and the Glasgow Site, a safe place to work.

1804—Powder mills constructed of heavy stone on 3 sides to direct potential explosions away from workers.

1850—Special boots for powderman to avoid sparks.

1882—Company used a horseshoe magnet to remove iron objects from powder before charging the grinding mill.

—Guards for moving parts of machinery such as belts, pulleys, gears, circular saws, jointers, and grinding wheels.

1895—Handrails for ramps and catwalks.

—Spiked feet on ladder bases.

1900—Rubber gloves, aprons, leggings, and special shoes introduced for handling chemicals or molten material.

1904—Water tubs provided for potential acid burns.

1912—First engineering equipment standard covered specification for rubber gloves.

1917—Goggles used for protection against chips and flying particles.

1932—Safety & Fire Division (S&F) standardizes requirements for safety equipment.

1955—"Buckle Up" for Safety campaign begins to promote use of seatbelts. Company salesmen began wearing them first.

1968—Glasgow Site installed Site fire alarm system.

1970—Glasgow Site built solvent storage building near 100 Building.

1980—Glasgow Site purchased emergency vehicle.

Statistics

As the company grew, the potential for injury grew as well. With more people, machines, and new technologies, the need became apparent to tabulate injuries so as to measure the effectiveness of safety programs. Statistics, when plotted on a graph since 1915, show that the safety programs have served us well and have indeed become a company and site TRADITION.

1815—First fatal accident—nine workmen killed.

1855—Wilmington Morning News wrote on March 23—"The number of persons killed in explosions has been reduced by the persistent efforts of the firm to discover and introduce safeguards to prevent catastrophe."

1911—Company began tabulating injuries for analysis and performance comparisons.

1912—43 lost time injuries per million manhours worked, company-wide.

1915—Injury reports submitted to Corporate headquarters for company-wide analysis.

1919—First *printed* list of plant standings arranged by cumulative severity rate.

1928—Company began program of safety awards for achievement.

1947—1 lost time injury per million manhours worked, company-wide.

1953—Off-job safety performance charted.

1969—Glasgow Site given its first Board of Director's Award after accumulating 1,564,000 exposure hours without a lost workday injury.

1980—Glasgow attains the number 1 position in the company for having accumulated the most exposure hours without a lost workday injury.

1981—Glasgow Site off-the-job injury performance better than company average for the first time since Site exceeded 1,000 employees in 1972.

—Glasgow regained the number 1 position for company-wide safety performance after having lost it to Sabine River.

1982—Less than 0.1 time lost injuries per million manhours worked company-wide.

1983—Glasgow's 13th BOD maintains our tradition of safety.

Occupational Health

Occupational Health programs have taken on many forms. From early medical services, through research at Haskell Laboratory, the TRADITION of employee safety is carried on. What TRADITIONS are we going to be remembered by? Will our efforts save someone from have a disabling injury? The TRADITION of Du Pont's safety program has brought safety to a level where we share the responsibility for the safety of everyone around us. *S*afety and *H*ealth *A*re the *R*esponsibility of *E*veryone.

1805—First doctor hired, Dr. Didier—$50 for 2 years service.

1907—First Aid Handbook published titled "Compilation of the Temporary Surgical Measure to be applied in Cases of Injury Pending the Arrival of the Physician."

1915—Periodic physical examinations for employees initiated.

1916—Construction of first separate medical facility completed at Parlin F&F Plant. Dr. Gehrman named medical director.

1925—Company initiated X-rays as part of employee physical examination.

1935—Haskell Lab established to study toxicological effects of chemicals for in-process use and sale.

1949—Procedure issued for disposal of fluorescent light tubes due to a beryllium coating causing a health hazard.

1954—Company-wide noise seminar held.

1960—Company began audiometric tests as part of employee physicals.

1977—Occupational Health group established as part of S&F Division and began auditing plants for industrial hygiene practices.

1979—Glasgow Site assigns full time Occupational Health Specialist.

1981—Glasgow Site initiated Personal Environmental Record System (PERS) for individual history of chemical exposures.

B

The Boss Speaks Out

Over the years, DuPont CEOs have written and spoken extensively on safety, as the following selections show.

Irénée du Pont
Chairman, Board of Directors (1926–40)

If there is a right way and a wrong way to do a given operation and the right way involves less opportunity for a careless move on the part of an employee, by getting him to adopt the safer course, we are doing safety work. By such methods, year by year, "accidents" will be decreased.

. . . the word "accident" is just a convenient designation of something that happens unintentionally with more or less undesirable results. Unfortunately most people think of "accidents" as occurrences which can be blamed on the

Almighty. . . . this is a serious error. It is becoming more and more apparent that everything which happens is the result of a previous cause—that a man cannot be injured in the sense of being involved in something not susceptible to human control.

12/1926

W. S. Carpenter, Jr. Chairman, Board of Directors (1948–62)

We in the DuPont Company long ago concluded that the safety of employees is of the greatest interest to management, ranking in importance with production, quality of product and costs. We have found that maintenance of safe operating procedures in our plants is of benefit far beyond any resulting dollar savings, the human values involved being of greater importance to both employer and community. Also, the acceptance and practice of fundamental safety principles by management and men, with the reduction of personal injuries to a minimum, inject an element of team play which does much to foster a spirit of friendly cooperation throughout our Company.

10/1946

I. S. Shapiro Chairman of the Board (1974–81)

When I signed on with DuPont years ago, I was given the usual assortment of pamphlets on pensions and insurance programs and that sort of thing, along with a couple of books on company history.

One thing in particular fascinated me at the time, and it still does, and that was the emphasis on safety. The message was loud and clear: Working safely is a condition of employment. Work safely—or don't work here at all. . . .

Now, everybody is in favor of safety, but I had never run into anything quite like DuPont. There were rules and regulations of every kind. Walk down the stairs and somebody would say, "Hold on to the handrail." Order a new file cabinet, and two fellows would show up to bolt it to the wall so it couldn't tip over. Sit in a chair and tip it back toward the wall, and somebody would say, "Keep all four legs of that chair on the floor."

All of this . . . in the Legal Department's offices. Until then it hadn't occurred to me that being a lawyer was such a risky occupation. . . .

In my initial reaction to DuPont, I turned the safety message around backwards. I assumed that the thing that produced the safety records was all those rules and regulations. Later . . . it dawned on me that these were not the cause but the effect. . . .

We find a fairly direct and nearly inviolable correlation between the safety record of a plant and its operating effectiveness. . . . We have convinced ourselves that safety does pay by making the entire operation of the DuPont Company more effective. There are benefits in terms of efficiency, of improved morale, and of reduced absenteeism. . . .

Safety in the industrial setting is primarily a responsibility of the private sector. That applies whatever the government does or doesn't do. It applies with or without OSHA. . . . Just because some practice conforms to a regulation does not mean it is safe.

5/1980

E. G. Jefferson
Chairman of the Board (1981–86)

Safety is good business in many ways. If you tour an operation that has demonstrated good, sustained safety performance, you will usually find the added benefits of good housekeeping, good product quality, and high morale. For the analysis and training that are essential to good safety bring also the benefits of superior operational control. The costs of injuries in terms of human suffering, the high costs of associated litigation, the costs of equipment damage and outage—all these can seriously threaten the health and reputation of the enterprise. Prevention is the only acceptable remedy. . . .

Safety provides us with a valuable corporate asset, one that brings much good will. Private-sector, multinational companies exist in the context of a social contract based on mutual benefit with the public and the governments in the countries in which they operate. To continue to manufacture in complex and potentially hazardous systems requires that we conduct ourselves in a manner that rewards the public's trust.

11/1986

Richard E. Heckert
Chairman, Board of Directors (1986–89)

At DuPont we have a long and proud tradition of setting for ourselves the highest safety standards. In recent years, pressures from outside have intensified, with new requirements and new criteria being imposed by government agencies. As a result, there may be a natural tendency to feel that the burden of responsiblity for maintaining a safe workplace has shifted outside the company. Of course that has not happened. We are obliged, as always, to live with our own performance.

One danger arises from the large amount of concern focused in recent years on health hazards in the workplace. These chronic hazards are much in the public spotlight, and they are often subtle and exotic. New knowledge is constantly coming from the laboratory and this requires us to devote time, money, and attention to learning how to manage these risks.

The danger is that this activity can distract us from doing our best to control the more immediate and obvious risks, the ones with which we have always been familiar. Their potential for harm is just as real as ever. The fact is . . . we are obliged to do our best on both. To put it bluntly, reducing exposure levels to parts per billion means very little if the plant blows up . . .

We have studied and planned, we have allocated new resources, and we have devised new administrative controls all for the purpose of improving our safety performance. But in the end, this effort is going to make a difference only if we, as individuals, are constantly dedicating ourselves to the task at hand.

5/1979

Edgar S. Woolard Jr. Chairman, Board of Directors (1989–)

Waste Reduction

We have to put behind us the notion that the inevitability of some waste generation therefore makes any waste acceptable—hazardous or not. We have to continue to design and implement waste minimization technologies and other waste reduction strategies. I want us to create a corporate culture in which there is no such thing as industrial waste. I believe anything that goes out the waste pipes may well be something that can be recycled, reused or sold. Today in our plants around the world, we recover nearly 1 billion pounds per year of polymers and polymer intermediates through waste minimization programs.

Zero Emissions

Another issue . . . is that of zero emissions. . . . DuPont's goal is zero emissions of carcinogens. In addition, industry may need to move more aggressively . . . with regard to other materials. . . . Zero emissions in an *absolute analytical* sense may not be possible. But for all *practical purposes* zero emissions, where necessary, may be achievable.

Respect for Local Communities and Natural Areas

The most serious environmental activism any company is likely to encounter is the grass-roots variety. Communities around the world are appropriately questioning health and safety considerations associated with . . . industrial plants in their midst. . . . Respect for local communities means involving residents in discussions about activities that potentially affect their safety, health and environment.

Natural systems have worth independent of their economic value. . . . Although land must be cleared, harvested, drilled, mined or otherwise disturbed for us to manufacture goods or reclaim resources, it does not follow that natural areas can be mistreated with impunity. There are ways of minimizing our impact and we must employ them.

Global Warming

For DuPont in particular, we have to pay close attention to the greenhouse properties of substitutes as we phase out CFCs (chlorofluorocarbons). [And] because we produce oil, gas, and coal, policies to slow the rate of buildup of greenhouse gases will affect us as a company. We want to be both part of the debate and part of the solution.

Employees are the key to corporate progress in addressing environmental concerns. They can exemplify the company's commitment by the way they do their jobs, and they are the company's ambassadors in the communities where they live. If they believe in the company, their neighbors probably will too. But if employees are skeptical or cynical about the company's commitment, nothing the executives say is likely to convince the public.

12/1989

C

Some Thoughts for the Safety Supervisor's Notebook

In this collection of notes, "the thing speaks for itself." It is a miscellany of lists, aphorisms, and folk wisdom that the authors have culled from long careers in the management trenches.

What Makes a Good Manager?

- Integrity—honest, sincere individual.
- Personal commitment to employees and good employee relations.
- Person who gets things done.
- Person who understands broad intents versus the rules—not hung up by details.
- Intelligence—quick mind.
- Person respected by those above and below.

- Open and able communicator; good listener.
- Nonpolitical person—calls a spade a spade; tells those who need to know when/what is wrong; doesn't let personal ambition show or get in the way.
- Bridge builder versus negativism; team builder—compromise versus confrontation.
- Person who has been in the trenches and understands needs—both exempt and nonexempt.

More on Managers

- Managers are appointed from above.
- Leaders are selected from below.
- Managers need to harness leaders.
- Managers do not try to change behavioral characteristics if they do not affect performance.
- No one motivates anyone else; you create a climate where individuals motivate themselves.

Managers Who Fail

- Cause is usually not lack of job knowledge or intelligence but personal characteristics and inability to get the job done through people.

Good Supervisory Traits

- Persuasiveness—one is decisive, can lead, understands others, and speaks and writes well.
- Intelligence—one is technically competent and has a quick mind.

- Energy—one is confident and ambitious and shows initiative.
- Stability—one is sincere, dependable, loyal, honest, and cooperative and works hard.

What to Do as a Supervisor

- Don't hoard authority—get more; give more away; accept more.
- Don't delegate:
  —Appraisal.
  —Discipline.
  —Promotion.
- Live up to commitments; be:
  —Predictable.
  —Positive.
  —Accurate.
- In appraising employees, remember:
  —Emphasis on promotion drains satisfaction from the present job.
  —Upgrading of skills and pride in work are very important to the organization.
  —Everyone wants to be somebody.
  —Avoid criticism.
- Be emotionally mature:
  —Avoid hasty action.
  —Sustain yourself on less praise than you deserve.

Good Employee Relations

- Know your people.
- Let them know what is expected.
- Keep them informed.
- Answer questions and resolve problems promptly.
- Achieve mutual confidence, respect, and trust.

What to Look for in Employees

- People who are:
  —Stable.
  —Persuasive.
  —Intelligent.
- People Who:
  —Want to work.
  —Work well with people.
  —Can cope with change.

Most Important Words

6. I admit I made a mistake.

5. You did a good job.

4. What is your opinion?

3. I need you.

2. Thank you.

1. We.

Sooner or Later[1]

- A wise individual discovers that life is a mixture of good days and bad, victory and defeat, give and take.
- One learns that it doesn't pay to be a too sensitive soul; that you should let some things go over your head like water off a duck's back.

---

[1]By Wilfred Peterson.

- One learns that the person who loses his or her temper usually loses out.
- One learns that all people have burnt toast for breakfast now and then and that you shouldn't take the person's grouch too seriously.
- One learns that the quickest way to become unpopular is to carry tales and gossip about others.
- One learns that buck-passing always turns out to be a boomerang and that it never pays.
- One comes to realize that the business could run along perfectly well without one's assistance.
- One learns that it doesn't matter so much who gets the credit so long as the business benefits.
- One learns that even the lowest-status employee is human, and that it does no harm to smile and say "Good morning" even if it's raining.
- One learns that most of the other people are as ambitious as oneself, that they have brains as good as or better than one's own, and that hard work, not cleverness, is the secret of success.
- One learns to sympathize with the youngster coming into the business, remembering the bewilderment most feel when starting out.
- One learns not to worry when a promotion is lost because experience has shown that if you give your best, your average will break pretty well.
- One learns that no one ever got to first base alone, and that it is only through cooperative efforts that we move on to better things.
- One learns that bosses are not monsters, trying to get the last ounce of work out of you for the least amount of pay, but that they are usually pretty good people who have succeeded through hard work and who want to do the right thing.
- One learns that folks are not any harder to get along with in one place than another, and that the "getting

alone" depends about ninety-eight percent on one's own behavior.

Try It Ben's Way[2]

I made it a rule to forbear all direct contradiction of the sentiment of others and all positive assertion of my own. I even forbade myself the use of every word and expression in the language that imparted a fixed opinion; such as "certainly," "undoubtedly," etc., and I adopted instead of them "I conceive," "I apprehend," or "I imagine a thing to be so and so . . ." or "It appears to be . . . at present." When another asserted something that I thought in error, I denied myself the pleasure of contradicting him abruptly, and of showing immediately some absurdity in his position, and in answering I began by observing that, in certain cases or circumstances, his opinion would be right but in the present case there appeared or seemed to me some difference, etc. I soon found the advantage of this change in my manners; the conversations I engaged in went more pleasantly. The modest way in which I proposed my opinions procured them a readier reception and less contradiction; I had less mortification when I was found to be in the wrong; and I more easily prevailed with others to give up their mistakes and join with me when I happened to be in the right.

Noah's Principle

- No praises for predicting rain—praise only for building arks.

[2]Passage is attributed to Benjamin Franklin.

A Good Creed for Difficult Times

Every morning in Africa, a gazelle wakes up. It knows it must run faster than the fastest lion, or it will be killed. Every morning a lion wakes up. It knows it must outrun the slowest gazelle, or it will starve to death. It doesn't matter whether you're a lion or a gazelle; when the sun comes up, you'd better be running.

D

Some How-To Cases

The following narratives fall largely in the realm of employee relations, but demonstrate the general principle that treating people properly and in a spirit of cooperative problem solving will help in achieving corporate goals, including those established for the safety program. In this atmosphere, it is easy to get the participation of employees at every level in reaching for a zero incidents goal.

Case #1: All-Out Cooperation

One of management's greatest challenges is to utilize fully the potential in the wage roll and lower management levels. Lately much effort has been put to doing so.

One successful approach follows:

- The site manager was told that the plant's output was not needed, but the plant could continue to operate if he could run a viable business on outside sales.
- The manager carried this message down through all levels, to reach every employee of the plant. He asked for suggestions and team effort.
- Restrictive work rules were totally eliminated. People started working together to get the job done.
- Wage roll employees with special talents were used to solve customer problems—including visits to customers' plants.
- Today, excellent morale and esprit de corps are evident, and the plant is operating successfully.

Case #2: Morale and Training

In one plant, the following measures were taken:

- The site manager spent about an hour every day informally talking with wage roll people as he walked around the plant.
- Supervisors were encouraged to ask and actively listen for group input to solve day-to-day problems.
- Group discussions of proposed changes were encouraged.
- During a plant open house and during a scheduled higher management visit, mechanical improvements were displayed and explained by the mechanics suggesting the improvements.
- Employees were kept informed of visitors and contractors, and of fellow employees who had experienced important events, including births in

the family. Possible changes affecting employees were communicated well in advance.

We have found that the wage roll group can be kept informed of changes, visitations, unusual occurrences, and safety problems through computers in control rooms, shops, and the cafeteria. Supervisors can assign one person in a computer group to accept and enter such data so that other individuals can, at their leisure, call up a particular package of such information accumulated over the preceding four weeks. This approach can have interesting results:

- In the spirit of togetherness, a mechanic offered to help his manager calculate his (the manager's) pension on the shop computer. Of course, the manager was not obliged to accept this offer.
- As a general principle we have found that employees often have suggestions for improvements and changes. The most effective reward for employees is the knowledge that their supervisor actively listens and aids them in getting the improvements established or helps them find even better ideas.
- We have found that employees usually don't crave greater responsibility. But they do want the recognition for their efforts that such information sharing generates.
- One particular use of computer access was in "playing" a multiple-choice computer game where the answers required detailed knowledge of safety rules. This turned out to be an excellent learning reenforcement tool. Before long, it was found that every mechanic knew every answer.

Case #3: Perceived Operations versus Actual Operations

One plant manager found that a review of the actual operation of a large chemical plant as opposed to the perceived operation was quite helpful. The manager assigned one top operator to work on the day shift until he had talked with every other operator and reviewed the standard operating procedures with each of them. This top operator was then encouraged to propose changes in the standard operating procedure to achieve consistency.

This procedure proved to be an excellent refresher training tool for each operator. It also produced an excellent operating manual.

Case #4: Developing Performance Standards

We found that establishing performance standards in the following way provided a great training tool.

In an organization with one shift supervisor and about five foremen per shift, the production superintendent asked the supervisor and the foreman to develop performance standards for production foremen under the four headings listed, using questions of the type shown:

1. Safety manual review
   - How should it be done? What standard should they meet?
2. Area patrol by supervisor
   - How often? How long? How detailed? What standard should they meet?
3. Housekeeping
   - What standard should they meet?
4. Operator training
   - What standard should they meet?

The superintendent found that group discussion to reach agreement was the best way of convincing some foremen who had been resistant to changes. For example, in one group one foreman said, "I patrol the area twice each shift." Another said, "I patrol about once in two days." Considerable "discussion" led to a consensus on the optimum procedure.

Principle: A group settling on its own performance standards with some supervisory guidance *but no dictation* develops standards useful not only for evaluation of performance but as a great training tool. Putting a listed standard under each heading as a guideline appeared to be helpful.

Case #5: Wage Roll Contributions to Improvement of Production and Safety Manuals

In a complex chemical plant one operator was given the task, on a day-shift assignment, of completely revamping and updating the safety manual. First he asked each operator how he could improve the manual. Surprisingly, in addition to suggestions on the usual safety rules and precautions, the operators indicated that they wanted detailed drawings of the internals of all vessels, with some brief explanation of how the unit should work. They also wanted diagrams of the control system, and several operators volunteered to help with drawings.

A top-quality manual was prepared.

Case #6: Long-Range Mechanical Supervisory Training

In order to solve a morale problem in the mechanical engineering group, a mechanic was assigned to work as helper to each senior mechanical engineer. This assignment was limited to one year, and it produced the following results:

- One mechanical engineer demonstrated that the assistance helped him produce three times as much work. He explained that half of his time was taken up informing or being informed, and his assigned mechanic did not have that problem.
- The mechanics developed more respect for engineers, and the engineers developed a whole lot more respect for mechanics.
- Because mechanics were picked with some care and promoted on finishing their assignments, we soon had some first-line supervision with much improved understanding of changes, including necessary safety approvals and the safety considerations required before the changes could be designed and adopted.

Principle: One-on-one training while "doing" is a great training method.

Case #7: Development of Site Policies and Acceptance of Change

Site policies on promotion, demotion, layoff, safety training, process training, equipment testing, and disciplinary actions all may be tentatively written first by top management, to be followed until revised. Then they may be reviewed and revised through each level of management. The wage roll should have input into the policies before final revisions.

This process was used to develop the detailed promotion, demotion, and layoff procedure at one plant. When a layoff of several people became necessary, there was little complaint because issues of shifting between groups, seniority rights, and so forth, had already been carefully considered.

Local management concluded that because changes had been suggested at every level and after careful consideration, no further training or explanation was needed. The actual policy development guided and finally accepted by management had accomplished many purposes, including maintaining and enhancing the morale of the group even in a period of adversity.

Case #8: Off-plant Activities Properly Handled Can Be a Morale Builder

At one plant shortly after start-up the employees asked to use a section of the plant for an employees' recreational association.

Management assured the employee group of its interest and said it would be happy to have a member of the plant staff help and advise individual groups.

The following volunteer groups were created:

- Swimming
- Baseball
- Fishing
- Dancing
- Golf

After several years of dedicated effort, including off-the-job supervisory input, and a helpful company contribution, this association now has:

- A large swimming pool, change house, etc.
- A golf course, though not the full 18 holes
- An active dance group
- A clubhouse large enough for an air-conditioned indoor tennis court
- A bar
- A picnic area
- Large barbecue pits and other facilities

For at least the first 40 years of its existence, this site never had a serious threat of a union election.

And this site has been a leader in safety, housekeeping, and productivity.

Case #9: The Written Record

One plant manager many years ago decreed that every personnel write-up should use the following format:

I told the man a, b, c. . . .
The man (by name) said a, b, c. . . .
We concluded that a, b, c. . . .
Signed:

This proved to be a good way to report what happened in an organized fashion but in a simple manner. *And it met legal requirements* to show that the employee was "told" specifically.

Bibliographical Note

As the opening pages of this book make clear, the story of DuPont safety is based on the personal working experience of the authors, who, in turn, have drawn on the collective experience of thousands of DuPont employees who have gone before them and others who have been their contemporaries and colleagues. Although hundreds of records came into play, they are almost totally the internal memoranda, letters, policy statements, and directives of day-to-day operations, and thus are not cited in a formal bibliography.

There has been no attempt to draw upon the previously published materials, histories, guidebooks, and so on, that scholars would feel obliged to consult and religiously document for other scholars who might follow in their wake. Even the dozen or so historical treatments of the du Pont family and the DuPont corporation only touch upon safety as a peripheral phenomenon, and most such histories are out of print.

For readers who do not have extensive experience working in the field of industrial safety, the National Safety Council can be helpful in recommending publications premised on a how-to approach.

About the Authors

William J. (Bill) Mottel retired in 1991 as director of safety and occupational health for the DuPont Company after 38 years of service. He was appointed to that position in 1986, the sixth person to occupy the position since it was established in 1914.

He is a member of the board of the National Safety Council and has served as its vice president of finance and head of the Finance Committee, and is now vice-chairman of the board.

In 1989–90, Mr. Mottel represented U.S. industry with the Organisation for Economic Cooperation and Development in workshops on the prevention of major safety incidents and in the development of a guidance document on injury and loss prevention.

Before becoming DuPont's safety director, Mottel was manager of the Savannah River Plant, which the company built for the Atomic Energy Commission and operated for the Commission and, later, for the U.S. Department of Energy. He also served as production manager and employee relations director for DuPont's petrochemicals operations, and as employee relations director for DuPont operations in Europe, the Middle East, and Africa. In retirement, he currently serves on the Environmental Advisory Council for Rohm and Haas Company.

On October 4, 1993, Mr. Mottel was inducted into the Safety and Health Hall of Fame International in a ceremony at the Sheraton Chicago Hotel.

Joseph F. (Joe) Long is a registered attorney and belongs to the Tennessee State Bar, having graduated from Nashville Law School in 1978 as Doctor of Jurisprudence. His speciality is patent law.

His career in law began after his retirement from DuPont in 1984. His DuPont career included assignments as research analytical chemist, technical engineer, production supervisor, chief chemist, technical superintendent, production superintendent, and manufacturing plant manager, in U.S. facilities and in a Mexican subsidiary.

He is currently engaged in consultation on biological remediation, and he holds a patent on methodology for reducing the corrosive effects of acid rain.

David E. Morrison retired from the DuPont Company in 1985 after 29 years spent in editorial work and as public affairs manager for Engineering and Central Research. Among his editorial assignments was the preparation of annual and quarterly corporate reports for five years. In retirement he continues to edit *Extensions*, a newsletter serving 70,000 DuPont pensioners and their survivors.

He also has served as chairman of the council of the pioneering Academy of Lifelong Learning affiliated with the University of Delaware, and he continues to teach literature and linguistics courses at that institution.

Index